CMOS Capacitive Sensors for Lab-on-Chip Applications

Analog Circuits and Signal Processing

For other titles published in this series, go to
www.springer.com/series/7381

Ebrahim Ghafar-Zadeh · Mohamad Sawan

CMOS Capacitive Sensors for Lab-on-Chip Applications

A Multidisciplinary Approach

Springer

Ebrahim Ghafar-Zadeh
Department of Electrical Engineering
Ecole Polytechnique de Montréal
Montreal QC
Canada
ebrahim.ghafar-zadeh@polymtl.ca

Mohamad Sawan
Department of Electrical Engineering
Ecole Polytechnique de Montréal
Montreal QC
Canada
mohamad.sawan@polymtl.ca

ISBN 978-94-007-3180-6 e-ISBN 978-90-481-3727-5
DOI 10.1007/978-90-481-3727-5
Springer Dordrecht Heidelberg London New York

Softcover reprint of the hardcover 1st edition 2010

Printed on acid-free paper

Springer is part of Springer Science+Business Media (www.springer.com)

Contents

Abbreviations

ADC	Analog to digital converter
Al	Aluminum
Al_2O_3	Aluminum oxide
APTS	3-Aminopropyltriethoxysilane
ATP	Adenosine triphosphate
Au	Gold
BCE	Bovine capillary endothelial
CAD	Computer aided design
CBCM	Charge based capacitance measuremnt
CCD	Charge coupled device
CE	Counter electrode
CGM	Continuous glucose monitoring
CMOS	Complementary metal–oxide–semiconductor
CVC	Charge to voltage converter
DAC	Digital to analog converter
DC	Direct Current
DEP	Dielectrophoresis
DFT	Discret Fourier transform
DNA	Deoxyribonucleic acid
DSP	Digital signal processing
DWFP	Direct-write fabrication process
DW	Deionised water
ΔC	Capacitance variation
E	Electrical field
ES	Electrochemical sensor
fF	Femto Farad
GLK	Glucokinase
HA	Hydroxyapatite
H_2SO_4	Acid sulphuric
H_2O_2	Hydrogen peroxide
HTS	Highly throughput screening
HIV	Human immunodeficiency virus
IC	Integrated circuit

ISFET	Ion-selected field effect transistors
Inc	Incorporation
LBCS	Lab-on-Chip based capacitive sensor
LoC	Lab-on-Chip
MEMS	Micro-electro-mechanical systems
mM	Milli mole per one liter
MBCS	MEMS based capacitive sensor
MOSFET	Metal oxide field effect transistor
mm	Micrometer
mm	Millimetre
Ni	Nickel
Ni^{++}	Nickel cation
NTA	Nitrilotriacetic acid
$NiCl_2$	Chloride nickel
NMOS	B-channel mosfet
OPAMP	Operational amplifier
O_2	Oxygen
PBS	Polybenzoxazole
PCB	Printed circuit board
PCR	Polymerase chain reaction
PDMS	Polydimethylsiloxane
PMOS	P channel mosfet
PoCT	Point-of-care testing
Pt	Platinum
RCDA	Rational cycle decoding algorithm
RE	Reference electrode
SAM	Self-assembled layer
SEM	Scanning Electron Microscope
SC	Switch capacitor
SiGe	Silicon germanium
SC-μSOFC	Single-chamber, micro-sized solid oxide fuel cells
SQNR	Signal-to-quantization noise ratio
TEMED	N,N,N′,N′-tetramethylethylenediamine
Ti	Titanium
TiN	Titanium nitride
TP	Transition point
UGB	Unity gain bandwidth
UV	Ultra violet
WE	Working electrode

Chapter 1
Introduction

1.1 Overview of Lab-on-Chip

Laboratory-on-Chip (LoC) is a multidisciplinary approach used for the miniaturization, integration and automation of biological assays or procedures in analytical chemistry [1–3]. Biology and chemistry are experimental sciences that are continuing to evolve and develop new protocols. Each protocol offers step-by-step laboratory instructions, lists of the necessary equipments and required biological and/or chemical substances [4–7]. A biological or chemical laboratory contains various pieces of equipment used for performing such protocols and, as shown in Fig. 1.1, the engineering aspect of LoC design is aiming to embed all these components in a single chip for single-purpose applications.

1.1.1 Main Objectives of LoC Systems

Several clear advantages of this technology over conventional approaches, including portability, full automation, ease of operation, low sample consumption and fast assays time, make LoC suitable for many applications including.

1.1.1.1 Highly Throughput Screening

To conduct an experiment, a researcher fills a well with the required biological or chemical analytes and keeps the sample in an incubator for some time to allowing the sample to react properly. Afterwards, any changes can be observed using a microscope. In order to quickly conduct millions of biochemical or pharmacological tests, the researchers will require an automated highly throughput screening (HTS) [8], comprised of a large array of wells, liquid handling devices (e.g., micro-channel, micropump and microvalves [9–11]), a fully controllable incubator and an integrated sensor array, along with the appropriate readout system.

E. Ghafar-Zadeh and M. Sawan, *CMOS Capacitive Sensors for Lab-on-Chip Applications: A Multidisciplinary Approach*, Analog Circuits and Signal Processing,
DOI 10.1007/978-90-481-3727-5_1, © Springer Science+Business Media B.V. 2010

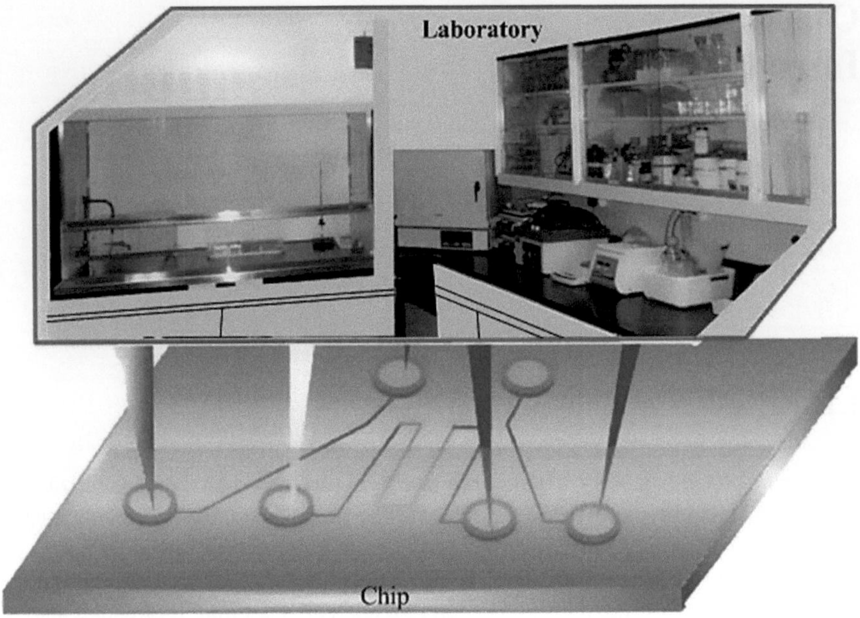

Fig. 1.1 From room scale laboratory to laboratory on chip

1.1.1.2 Early Detection of Disease

To find health problems before symptoms appear is known as early disease diagnosis [12]. Early genetic disease detection, such as in some forms of cancer, may prevent symptom development through early medical intervention [13]. Also, the identification of the infectious agent in its early stage of development is critical to limit the spread of deadly disease such as Human Immunodeficiency Virus (HIV) and to speed patients' treatment [14, 15]. For this, frequent testing for such diseases through cost-effective and easy-to-use LoC is the best solution.

1.1.1.3 Point-of-Care (PoC) Testing

PoCT refers to all diagnostic tests performed close to patient. For instance, bacterial infection and heart failure can be signalled by specific biomarkers in the blood [16–19]. Testing for such biomarkers can be carried out using portable devices in a doctor's office or even at home. In fact, PoCT has emerged as a key step in improving quality of health care, because in all cases, prevention is better than a cure.

1.1.1.4 Environmental Assessment

Air and water pollution is responsible for several health problems such as heart disease, lung cancer and microbial disease [20, 21]. The main objective of such monitoring and assessment is to determine the microbial agents found in water supplies, as well as the major air pollutants, including sulphur oxide, nitrogen oxide, carbon monoxide, volatile organic components and particulate matter [22]. Portable LoCs can effectively be functionalized to sense chemical gases in air or harmful microbial growth and contamination in food and water supplies [23, 24].

Now that you've been given a brief introduction to the LoC, let us now go deeper into the biological protocols and micro-technological solutions used in their miniaturizations.

1.2 From Macro to Micro Bioassays

As mentioned before, biology and chemistry are experimental sciences that are continuing to evolve and develop new protocols. Each protocol contains step-by-step laboratory instructions, lists of necessary equipment and required biological and/or chemical supplies. For instance, a widely applicable protocol for DNA extraction from blood is so noted in [25]: "(1) To 10 mL whole blood add 30 mL lysis buffer in a test tube, (2) shake the tube gently and incubate for 30 min at low temperature, and (3) centrifuge at 1200 rpm for 10 min at 4°C, and …". As shown in Fig. 1.2a, a certain volume of blood is sampled and mixed with a chemical solution, called lysis buffer, in order to break the cell membrane prior to DNA extraction. The mixture is incubated and then centrifuged at a specific speed and temperature. As seen in Fig. 1.2b, laboratory equipment, including a centrifuge, incubator and glassware, are routinely found in a sample preparation procedure. The miniaturized version of this assay is also shown in this figure. DNA amplification, fragmentation and fluorescent detection can also be performed through a polymerase chain reaction (PCR) system using a gel electrophoresis device and a DNA microarray [26, 27]. The miniaturization of this equipment is the main objective of LoC design.

1.2.1 Micro-scale Liquid Handling

Microfluidics is a key component for the automation of bioassays [28–30]. In fact, instead of glassware (e.g., test tube, pipette) for liquid handling, a network of microchannels, along with other microfluidic components including mixers, valves and pumps, are employed to direct, mix and control the fluids in a LoC platform.

The current microfluid technology is capable of precisely implementing the microfluidic/nanofluidic components through a variety of materials, including silicon, glass and polymers [31–33], and techniques using conventional microelec-

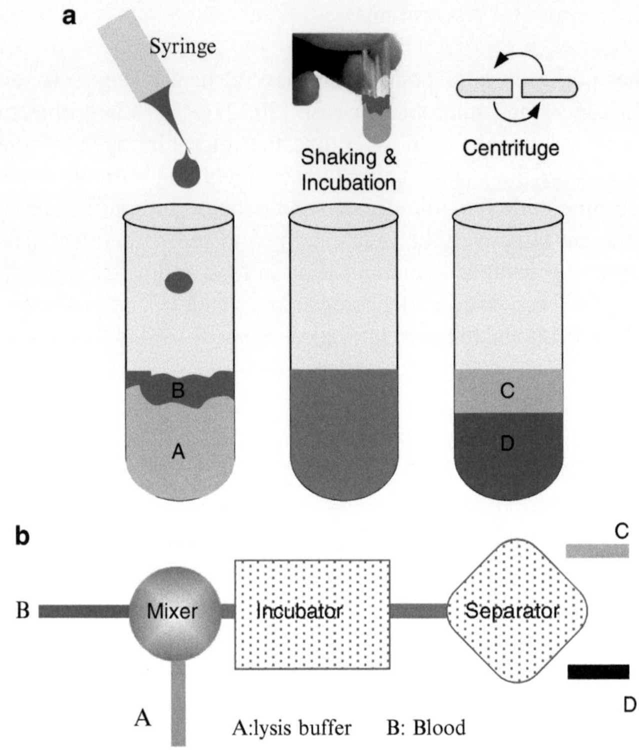

Fig. 1.2 A simplified diagram for miniaturization of biological protocol: (**a**) traditional methods and (**b**) integrated system

tromechanical system (MEMS) procedures or soft lithographical methods (e.g. hot embossing [34, 35]). Figure 1.3 shows some microfluidic chips implemented by Micronit [36]. Several functions, including microchamber assay, mixing and electrophoresis, is performed using these chips.

1.2.2 Thermal Management in Microenvironment

Temperature is the dominate factor in many biological activities or chemical reactions [37, 38]. For this reason, the incubation of a sample at a specific temperature is critical in biological and chemical processes. Likewise, the miniaturization of an incubator with precision control of temperature is an important task in a LoC platform. A micro-incubator consists of a microfluidic chamber, a micro-hotplate to generate heat, and microelectronic circuitry to control the temperature. Figure 1.4a–c shows a traditional CO_2 incubator (Cole-Parmer Inc.), a new small closed perfusion incubator chamber (LU-CPC-CEH, Harvard Apparatus Inc. [39]) and an on-chip

Fig. 1.3 Microfluidic chips fabricated by Micronit Inc (*micronews*, jun 27th, 2008, microfuidics)

incubator [40] fabricated through semiconductor technology with polymeric packaging, respectively. It should be mentioned the many incubators also regulate oxygen and carbon dioxide when incubating living cells samples [41].

1.2.3 DNA Amplification

A widely used biological protocol that requires a unit to control the temperature is the polymerase chain reaction (PCR) [37, 42]. PCR is employed to amplify the fragments of DNA molecules by enzymatic reactions. In PCR, DNA molecules are made by experiencing a series of temperature changes ranging from 50°C to 98°C. Figure 1.5a shows a thermal cycler device used for PCR experiments that has been commercialized by eppendrof Inc. [43]. Figure 1.5b shows a miniaturized PCR device innovated by Institute of Bioengineering and Nanotechnology (IBN) [44]. In this figure, four different locations (H1–H4), with different temperatures, is implemented in a microfluidic chip. This tiny device can rapidly prepare, purify and genetically analyse blood (red droplet in Fig. 1.5b). Magnetic nanoparticles are attached to particular cells in the sample then the cells are magnetically manipulated from one station on the chip to another station.

1.2.4 Sample Handling

A centrifuge is routinely used in a biological laboratory to separate particles of various sizes or weights. The principle operation of this device is based on centripetal forces applied on the suspended particles as seen in Fig. 1.3. The miniaturization of

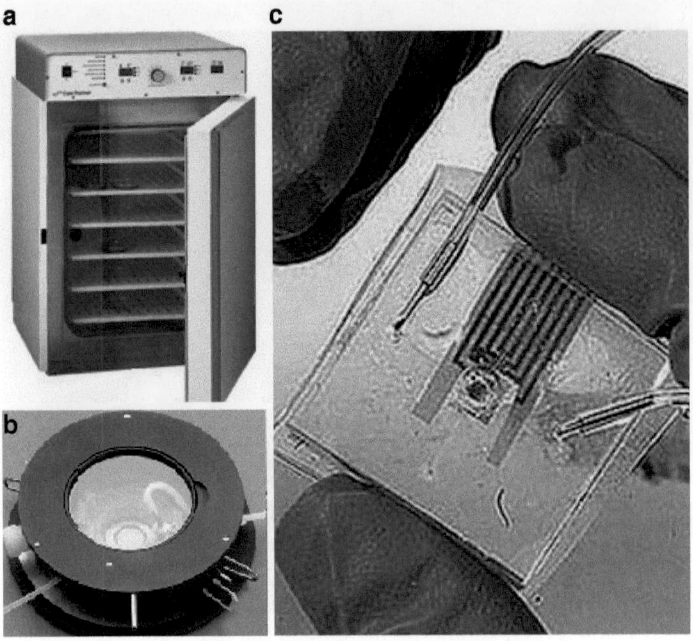

Fig. 1.4 From macro to micro incubator: (**a**) Traditional ("CO2 incubator" from Cole-Parmer Inc.) (**b**) millimetre ("LU-CPC-CEH" from Harvard Apparatus Inc.) and (**c**) microscale incubator

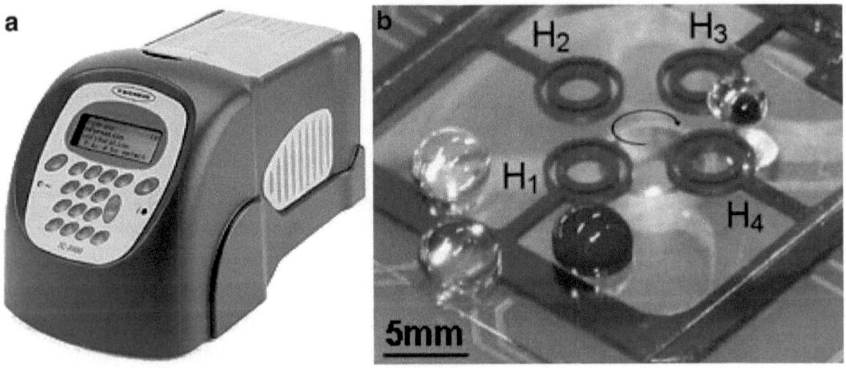

Fig. 1.5 PCR device: (**a**) Current ("PCR" eppendrof Inc) and (**b**) Future technology (News, RSC, 9 May 2008, "All-in-one gene detection on a chip")

such a system requires complicated microelectromechanical system (MEMS) procedures. In preference, the manipulation of particles is performed by LoCs through a biophysics phenomenon called dielectrophoresis (DEP) [45]. By applying a sinusoidal voltage in between two microelectrodes, a non-uniform electric field is generated to push the particles. The amount of applied forces on each particle depends on the electrical properties of medium, the frequency and amplitude of the applied voltage, the shape and size of the particle and the geometry of the microelectrode. To date, several papers reported DEP based devices for the manipulation [46], rotation [47], levitation [48] and separation [49, 50] of single cells. Among these works, Gascoyne et al. proposed a programmable DEP based system for sample handling purposes [51]. Their system is capable of detecting, fractionating, isolating and concentrating specific cells for cell lysis. Prior to cell lysis, cell trapping and focusing is performed on a spiral DEP electrode, as shown in Fig. 1.6a. A macro sale centrifuge is illustrated in Fig. 1.6b, and includes a rotating disk that can hold 12 test tubes simultaneously.

The applied electric field on charged molecules, such as DNA, results in electrophoresis phenomena [52]. Based on these biophysics phenomena, the charged molecules will move in the direction of electric field with different speeds corresponding to their sizes. Figure 1.7a shows a commercially available low cost device for electrophoresis (Topac Inc.). A high voltage must be applied to generate the required electric field in order to move the molecules through a gel substance. In a porous, sponge-like gel (e.g. hydrogel [53]), smaller molecules move more easily than larger ones therefore, a spectrum of molecules with different sizes is created. The new microscale electrophoresis neither requires this gel material, nor a high voltage power supply. Figure 1.7b shows a microfluidic chip (LabChip®) fabricated by Caliper Inc [54] for capillary electrophoresis.

Many biological procedures, such as optical detection of DNA molecules using fluoresce labelling techniques [55] or magnetic separation methods using Dynabeads® [56], have recently been miniaturized through micro- and nanofabrication techniques. This book is not intended to provide comprehensive coverage of bioassay, nor is it intended to tell you everything there is to know about miniaturization techniques. The focus of this section is to describe how to move from macro scale biological laboratory toward lab-on-chip. Let us wrap up this discussion by mentioning the main advantages of biological miniaturization.

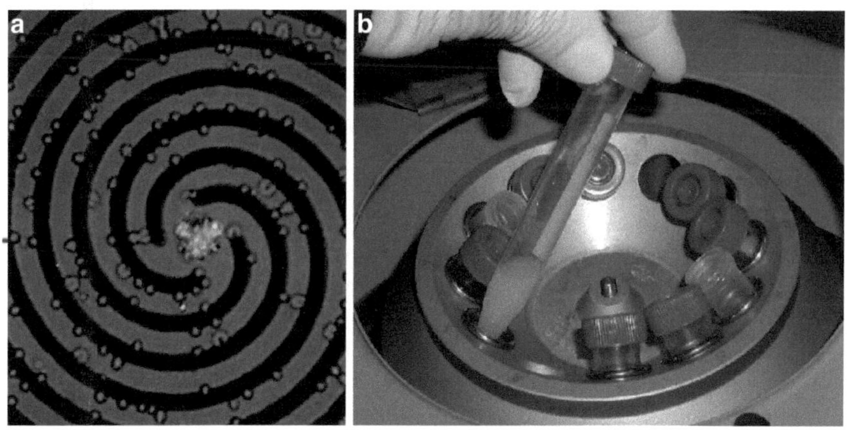

Fig. 1.6 Micro (**a**) Proceedings of IEEE, 2004, v. 92, n.1 (Fig. 11) and macro (**b**) scale centrifuge device

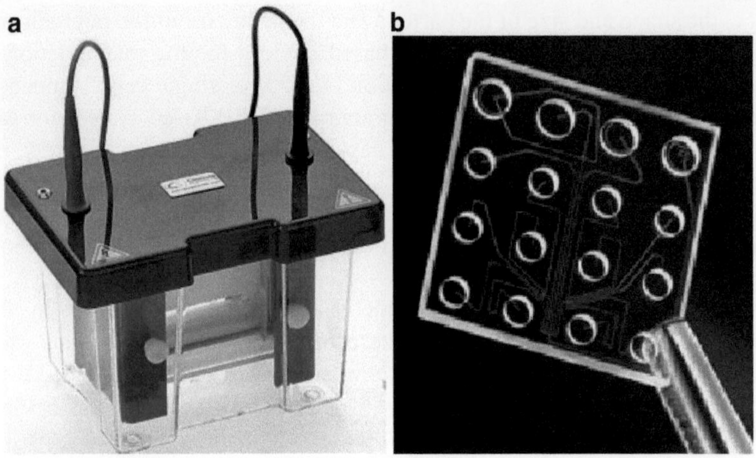

Fig. 1.7 Macro and micro tools for electrophoresis: (**a**) traditional tool (Topac Inc.) and (**b**) microfluidic based chip (Caliper Inc)

1.2.5 Advantages of Performing Bioassays in Microscale

In the beginning of this section, the current objectives of LoC technology were briefly described. A LoC system dedicated for these purposes allows several further advantages, including:

- Low voltage dielectrophoresis/electrophoresis: Miniaturizing the spaces between the electrodes increases the electric field intensity and consequently increases the force on the bio-particles within this field. Therefore, by applying a few volts instead of a few hundred volts, the required electric field intensity can easily be generated between the electrodes.
- Prevention of evaporation: Evaporation is a problem in any biological assay; this phenomenon suppresses the experimental results, especially where the sample volume is very low. A closed loop microfluidc network, in a LoC system, can prevent the sample from evaporation.
- Low sample consumption: Decreasing the volume of a sample contained in a miniaturized bioassay makes LoC systems attractive for work involving rare or expensive biological substances.
- Cheap and hand-held devices: Small devices are portable and can be placed in constrained spaces. As already mentioned, this portability is critical for PoC use. Such small devices can be fabricated using micro-fabrication batch processing which results in low cost products.
- Single cell experiment: Shrinking devices enable single-cell analysis. In other words, micro-fabrication technology allows for the implementation of tiny sensors the size of single cells or smaller. These small devices are necessary when performing analysis such as single cell recording or monitoring [57].

- Fully automation bioassay: As already mentioned, a complete bioassay consists of a series of functions that should be performed individually using different chips. A microfluidic structure is involved in the handling of the bioparticles (or chemicals) from one chip to another. A hybrid microfluidic/micrpelectronic system can be implemented to automate the biological assay containing a small sample.

1.3 CMOS-Based LoC

Standard CMOS technology offers the advantage of cost effective integrated circuits, programmability and control, embedded sensors and/or actuators and, as such, is a good fit for implementation of some of the essential functions of LoCs.

A holistic system-based approach to CMOS LoC design seamlessly integrates microfluidics, microelectronics and biochemical reagents and reactions on a miniaturized platform. As shown in Fig. 1.8, a micro-fluidic structure is required to direct the fluids toward sensing/actuating sites while the biochemical reagents are employed in the channel above the microelectronic devices. These devices are responsible to detect or to apply an electric signal for sensing or actuating purposes, respectively. Indeed, microelectronic CMOS technology allows one to fabricate an active substrate with different types of sensing or actuating sites that can be addressed in a random access mode by means of decoders. Photodiodes [58, 59], ion-selected field effect transistors (ISFET) [60], lateral/substrate bipolar transistors [61], and microelectrodes, realized in the topmost metal layer, are the front-end devices in CMOS process that can be employed for optical sensing, charge detection, temperature measurement and electrochemical sensing purposes, respectively. Additionally, the topmost metal layer can be patterned so as to become DEP microelectrodes or magnetic microcoils [62, 63].

Table 1.1 shows a summary of recent advances of CMOS-based LoC. As shown in this table, many biological and chemical applications, including DNA detection,

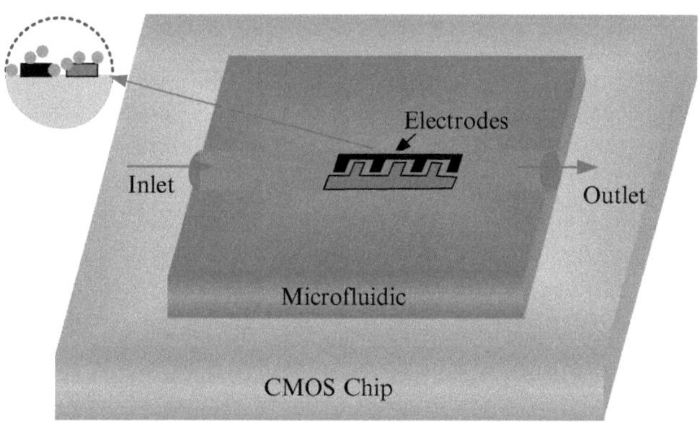

Fig. 1.8 A simplified diagram of a CMOS-based LoC

antibody-antigen recognition, DEP and magnetic manipulation, organic solvent and pH sensors can be performed through CMOS processes. As already mentioned, the sensing devices realized on CMOS chip are employed to detect the minute optical, thermal or electrical changes and readout the required information for further signal processing or monitoring purposes. More details of CMOS based methods to develop a LoC are provided in the following sections.

1.3.1 Manipulation Methods

By taking the advantage of DEP manipulation, an array of microelectrodes can be implemented using a CMOS process in order to generate a non-uniform field [62, 77–80]. Figure 1.9 shows a closed DEP cage above each electrode where a sinusoidal signal is applied in between the associated electrode and a conductive glass. By changing the electrical voltage of electrodes, it can be possible to trap a single cell or a cluster of cells and move it in the desired trajectory for biological assay purposes. Among the efforts made towards on-chip DEP manipulation is Manaresi et al., who developed a 0.35-μm CMOS chip comprised of more than 10,000 electrodes that enabled the displacement of living cells through software-controlled logic [81]. In this design, the standard passivation was used without any further etching or thin layer deposition processes. Since DEP force is proportional to the square of applied voltage, a high-voltage CMOS technology, with a large supply voltage, can effectively be used with fewer limitations on these forces. This issue was demonstrated by Current et al through 130-V 1.0-mum SOI CMOS fabrication technology for digital microfluidics [82]. The ability to handle, merge and divide the single droplets using DEP forces has recently emerged as a powerful tool for LoC applications. Also, high-voltage CMOS processes are a suitable for molecular fractionation applications, through electrophoresis techniques [83, 84]. Based on these phenomena, the ionically charged particles such as DNA molecules with different lengths can be separated in an electric field.

Magnetic manipulation of beads coated with biological material has been widely employed for several applications, including the diagnosis of infectious diseases [85, 86]. In such an application, a large group of functionalized beads are randomly spread in the analyte using a non-uniform magnetic field. The manipulation of individual beads offers a powerful tool for biological studies. An array of micro-coils can be realized using a CMOS process similar to microelectrode as shown in Fig. 1.10a. An interface circuit is also designed for independent current control of each coil [66]. Lee et al. demonstrated this CMOS-based manipulation method using BCE cells [87, 88]. These cells contain multiple magnetic beads formed using a natural engulfing process. In this design, three metal layers (M1, M2 and M3) were used to form three coils connected in series via a CMOS process, as shown in Fig. 1.10b. These coils were required to produce a high magnetic intensity at the center of the coils, above chip.

Table 1.1 A summary of CMOS based LoCs

Application	Front-end device	Process (μm)	Year	Ref.
DEP manipulation	Microelectrode Array	0.35	2004	[62]
Magnetic manipulation	Microcoil array	0.5	2006	[63]
DNA detection	Impedometric sensor	1.5	2005	[64]
Glucose monitoring	Amperometric technique	0.5	2007	[65]
Organic solvent detection	Capacitive sensor	0.18	2007	[66]
pH sensor	ISFET	0.18	2007	[67]
Antibody-antigen recognition	Optical method	0.18	2007	[68]
Bio-luminance detection	Photodiode array	0.5	2006	[69]
Antibody-antigen recognition	Capacitive sensor	1.5	2005	[70]
Bacteria growth monitoring	Capacitive sensor	0.18	2009	[71]
DNA detection	Capacitive sensor	0.18	2006	[72]
Toxic gas detection	Capacitive sensor	1.2	2002	[73]
Cell localization	Capacitive sensor	0.35	2004	[74]
Cell monitoring	Capacitive sensor	0.5	2007	[75]
Protein biosensor	Amperometric technique	0.5	2009	[76]

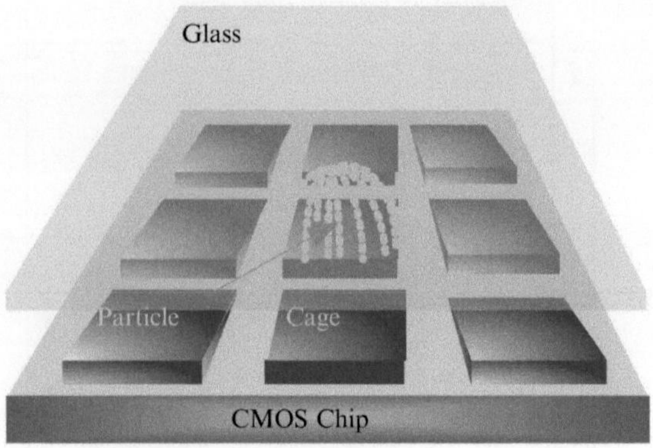

Fig. 1.9 Illustration of DEP microelectrodes on CMOS

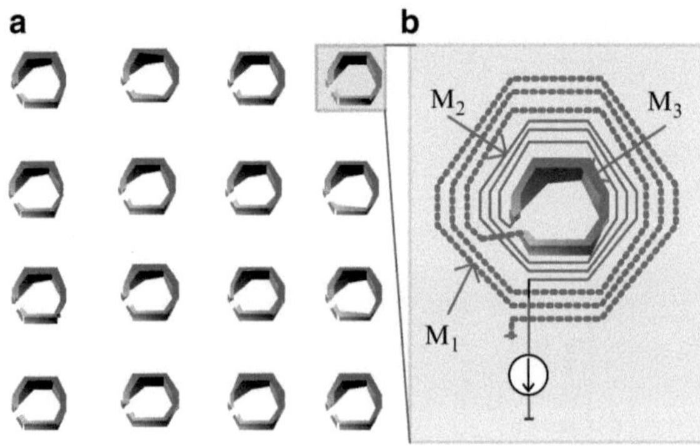

Fig. 1.10 On-chip magnetic manipulation: (**a**) microscopic image of chip containing an array of microcoils and (**b**) illustrations of a magnetic coil containing three metal layers

1.3.2 Optical Techniques

The changes in optical radiation impinging on the photodiodes embedded in the CMOS senor are detected by an integrated circuit. These optical signals can be emitted from different sources including fluorescent tagged particles [89, 90], bioluminescence assays [69] or an optical opening indicating the presence or absence of particles [74] as shown in Fig. 1.11a–c.

A CMOS chip with a large array of photodiodes offers a high resolution, low power consumption and highly integrated imager over conventional charge coupled device (CCD). The focus of current research on CMOS optical sensors is the implementation of hybrid systems consisting of optical filters (e.g., fluorescent filter) and integrated circuitry, as well as bio-recognition layers. In this way, Medoro et al. developed an integrated CMOS chip containing an imager using the aforementioned

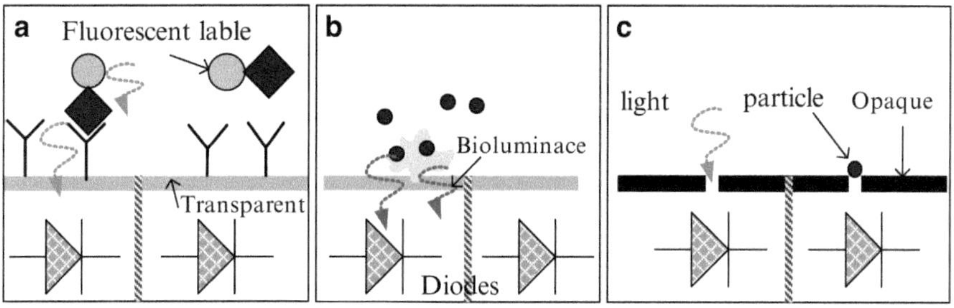

Fig. 1.11 Optical techniques: (**a**) fluorescent label, (**b**) bio-luminance and (**c**) opening

on-chip DEP manipulation method [62]. As shown in Fig. 1.1c, the passivated topmost metal layer in this process has been employed as DEP microelectrodes. This metal layer is also used as an opaque layer with the openings above photodiodes. Similar methodology was applied by Hartley et al. to implement an active pixel sensor for on-chip cytometry [91]. The optical detection of transparent cells can be modified by reducing the effect of optical noise and consequently increasing the sensitivity. However, further biochemical procedures such as fluorescent or bioluminescent assays should be performed to improve the selectivity.

The luminance emitted from certain chemical reactions can be used to detect bioparticles such as pathogens and proteins [92, 93]. Taking advantage of such a biosensing method, Eltoukhy et al. reported a CMOS sensor comprised of an 8 × 16 pixel array of $P^+/N/P_{sub}$ photodiodes, along with analog to digital converter (ADC) and DSP system [69]. Such photodiodes have a very low dark current and junction capacitance and consequently improve the overall performance of the imager. Also, further circuit techniques have already been employed to improve the noise factor of such image sensor arrays [94–96]. Figure 1.12 shows a pseudo-differential pixel. This interface circuit is used to decrease the readout noise level of optical signal detected by photodiode where BIAS, VREF, VSET and RESET voltages are applied as bias, reference or control signals respectively.

Fluorescent labelling assay is widely used in biochemistry as a selective optical detector for diagnostic and research purposes. Through a fluorescent assay, the fluorescent labels (e.g., SYBR-Green [97]) are attached to bioparticles in analyte. By exposing the functionalized sensing layer (e.g., DNA probes or antibodies) only the target molecules/cells are trapped on the surface of the sensor, while other non-bonded molecules are washed out. Therefore, the fluorescent label of target molecules can be detected using an optical sensor. A fluorescent exciter along with a fluorescent filter is usually needed to properly sense the light that reaches the photodiode from the label. Of course, through novel, filter-less CMOS techniques [90], less complexity is required for on-chip fluorescence detection. However, other label-free sensing methods, such as electrochemical methods, have recently attracted attentions for incorporation in hand-held LoC designs. Figure 1.13a shows a very advanced Genchip fabricated by Affymatrix Inc. for DNA detection. Figure 1.13b shows a large array of DNA probes in Genchip coupled with labelled molecules.

An optical scan and further image processing should be performed to record the desired information for genetic diagnosis applications. The main objective of the miniaturization of fluorescent labelling assay is to fabricate a CMOS imager, integrated with a genchip for DNA detection, to be used in a handheld device.

1.3.3 Electrochemical Sensors

These sensors rely on charge transfer from an electrode to the target sample. By applying a DC voltage on an electrode, the relationship between current and voltage is obtained as a function of the chemical/biological properties of a sample. This simple measurement method is called voltometry [98]. In a different configuration, the impedance in between the electrode can also measured for sensing purposes.

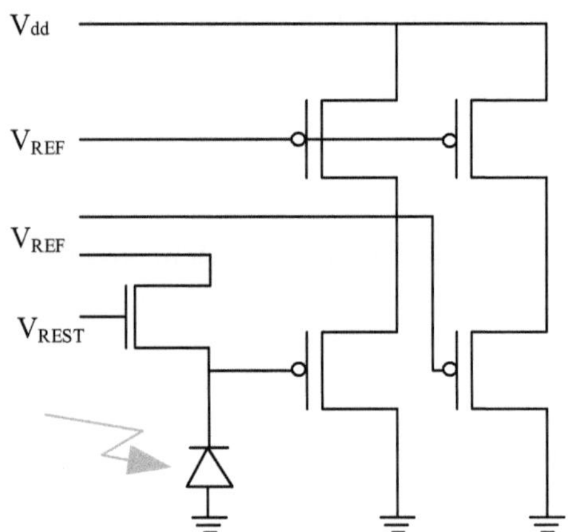

Fig. 1.12 Schematic of an active pixel for a CMOS optical sensor

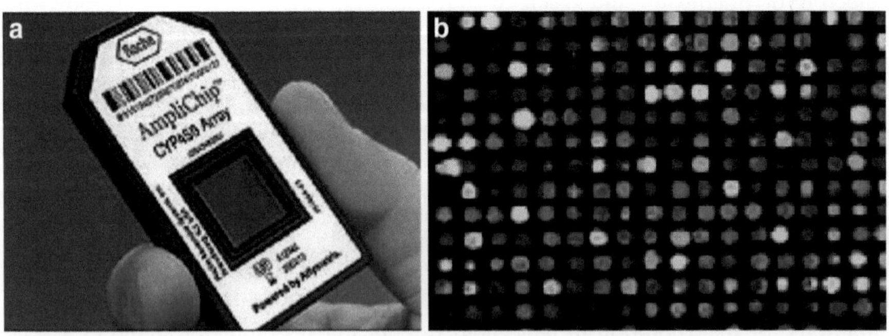

Fig. 1.13 An advanced biochip for DNA diagnosis: (**a**) GenChip implemented by (Affymatrix Inc.) and (**b**) the scanned optical data from Genchip

Conventionally, an electrochemical sensor (ES) consists of three electrodes – a working electrode (WE), a counter electrode (CE) and a reference electrode (RE), rather than two signal and ground electrodes, in order to avoid the polarization problem seen with ground electrodes. The integration of the required circuitry for impedance measurements on the same chip of microelectrodes can significantly decrease the need for references and the effect of parasitic capacitance that arises from the interconnection of electrodes and the measurement system. Hassibi et al. have demonstrated this key issue in 0.18 CMOS process using Al/1%Si metal layer for WE, CE and RE, as shown in Fig. 1.14 [64]. It should be mentioned that aluminum is not as widely used in biochemistry as gold and it is covered by an oxidation layer under normal environment conditions. However, considering that this work focuses on integrated circuit design, the specification of the proposed ES array makes it suitable for many biochemical detection purposes, including those used to detect DNA and proteins [99, 100]. In this way, Levine et al. reported another integrated impedometric ES array implemented through a 0.25μm CMOS process for DNA detection [101]. Further post-processing steps including etching and sputtering procedures were performed to replace the gold electrodes instead of standard aluminum electrodes.

The Ion-Selected Field Effect Transistor (ISFET) is a fully electronic device for direct sensing of charges. An ISFET is, in principle, a MOS transistor with the gate metal replaced by a solution under test. As shown in Fig. 1.15, the ionically charged molecules/cells in close proximity of insulation layer above the gate affect the depth and consequently the resistance of the channel (R_L). On other words, the drain current, which depends on the gate-source voltage, varies as a function of the number of charged particles trapped on the surface of the insulation layer.

A recognition layer, such as DNA probes [102] or antibodies [103] can be immobilized to catch these charged particles. A well-known application for ISFET sensors is a pH monitoring [104]. pH, which is an indication of activity of solution, is defined as the relative amount hydrogen ion activity and is expressed on a logarithmic scale. The miniaturization of a pH sensor is very important for many applications, including the metabolism of cellular activities [105, 106].

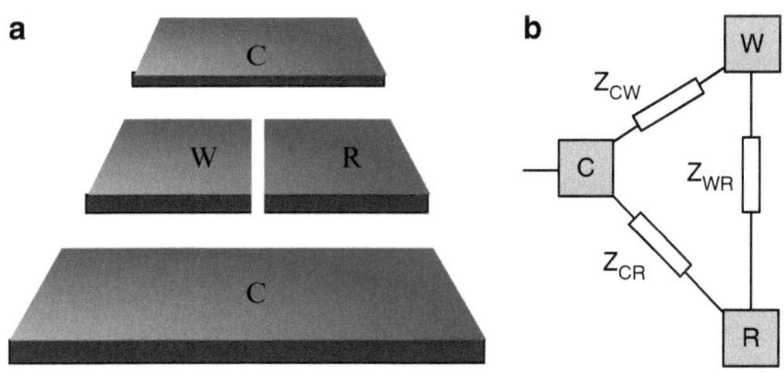

Fig. 1.14 On-chip electrochemical sensor: (**a**) SEM image of an electrode atop a CMOS chip and (**b**) the electrical model of each electrochemical cell

Among the efforts made to integrate an ISFET for this purpose, Lauwerse et al. reported a 1.2 μm CMOS chip for multiple sensing purposes including pH measurement [107]. In another effort, Hammond et al. reported a single chip sensor using a standard 0.6 μm CMOS process [108]. A packaging technique was applied to prevent the access of ions to places other than the active area above chip. This is because; the presence of ions can increase parasitic capacitances and resistances and consequently decrease the performance and sensitivity of the sensor. Also, the precision of ISFET is always affected by temperature variations. Thanks to the design and implementation of highly precision circuitries, a temperature measurement system is integrated with the ISFET sensor to compensate for the resulting errors [109].

1.3.4 Mechanical Sensors

These sensors function according to piezoelectric phenomena, defined as the coupling between electrical and mechanical effects [110]. By exciting some material, such as quartz or zinc oxide, with a sinusoidal voltage, a mechanical deformation is achieved [111]. In turn, this mechanical deformation results in a charge displacement and the generation of a measurable electric charge. The resonance frequency (f_0) of this electro-mechanic system can vary as a function of surface mass loading ($\Delta m/A$), as published for the first time by Sauerbery in 1959 [112] and shown in Eqn. 1.1.

$$\Delta f = C f_o (\Delta m / A) \tag{1.1}$$

where C is a constant and Δf represents the variation of f0. Based on this relation, several biochemical sensors have been developed to measure the frequency shifting as a function of mass. This method of measurement is the principle behind several desktop sensors, including the Quartz Crystal Microbalance with Dissipation monitoring (QCM-D), commercialized by Q-Sense Inc. [113].

Until now, many efforts have been made to implement the miniaturized piezoelectric based sensors through standard technologies relying on the piezoelectric property of GaAs [98]. As silicon is not piezoelectric, a fully integrated system can not be implemented only using CMOS processes. For this reason, further post-processing is usually required to realize a piezoelectric membrane; a top layer of ZnO on a CMOS chip is one such example [114, 115]. Among the mechanical sensors implemented through CMOS processes, resonating cantilevers have attracted attention for its wide range of applications, including DNA detection and antibody-antigen recognition [116–120]. Figure 1.16 shows an illustration of a cantilever fabricated using a CMOS process.

As seen in this figure, the structure was realized using a back-etching method, where the polymeric recognition element is coated on the tip of the cantilever. As the SiN and SiO have different thermal expansion coefficients, the cantilever can be deflected by the heat generated by polysilicon heater. An external magnetic field along with a micro-coil, which acts as the heater, can be employed as a magnetic actuator in the cantilever. A Wheatstone bridge, including four transistors, is used to monitor deflection (Fig. 1.17). As shown in Fig. 1.17, the closed loop system, with two transfer functions, H1(s) and H2(s), is implemented using analogue building blocks, including low-pass filters and amplifiers. It should be mentioned that R is representative of the resistive property of the actuator, whether from the external magnetic coil or the on-chip heater. Also, f_0 in this system is a function of applied force on the tip of the cantilever due to the interaction between the recognition layer and the analyte, or the extra mass deposited on the cantilever.

1.3.5 Magnetic Sensor

Magnetic fields can simply be detected using resonance or Hall-effect methods [121, 122]. Magnetic Hall sensors are the most widely used integrated magnetic sensors for many applications including position, angular or current measurements [123].

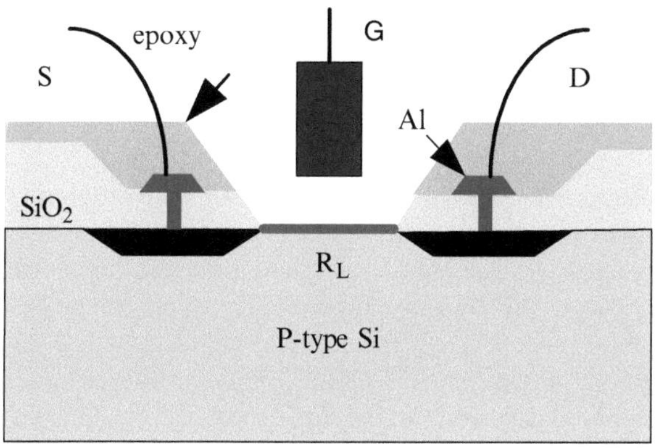

Fig. 1.15 Simplified diagram of an ISEFT device

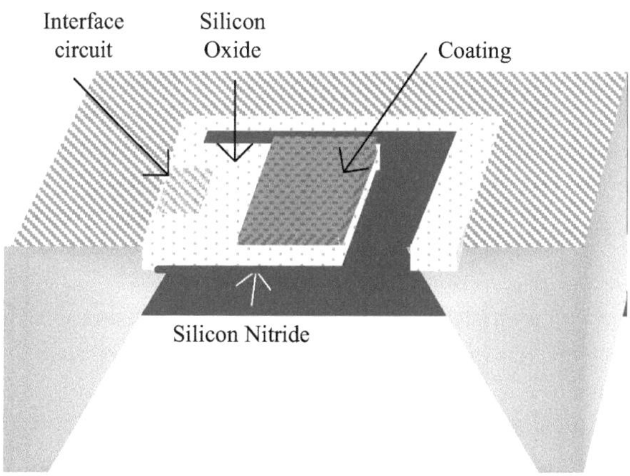

Fig. 1.16 CMOS-based cantilever

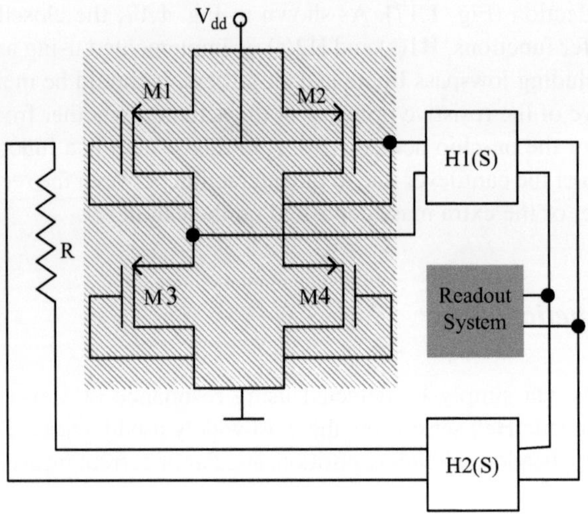

Fig. 1.17 Schematic of an interface circuit for a cantilever

Figure 1.18 illustrates the principle of a Hall sensor. The differential voltage (V_{AB}) is measured where I and B are the applied current and magnetic flux density, respectively. As already mentioned in Section 1.2.4, the functionalized magnetic beads can be employed for sensing purposes. These beads can be detected using integrated magnetic sensors. Until now several papers reported the integration of Hall-effect magnetic sensor using a CMOS process, but less attention has been paid on CMOS-based sensors for magnetic bead bioassay. Among these few works, Ayture et al. reported a CMOS magnetic sensor for infectious disease diagnostics [124, 125]. Figure 1.19a, c shows the proposed CMOS magnetic system used for this purpose. The external magnetic field pushes the functionalized magnetic beads toward sensing sites. After enough time has passed for the biparticles on the beads to bind on the substrate, magnetic field of opposite direction is applied to wash out the non-bounded beads. Therefore, a highly accurate magnetic sensor is realized accordingly.

1.3.6 Temperature Control

A CMOS system for temperature control consists of sensing devices and heating elements. The most straightforward way of measuring temperature is to take advantage of either parasitic bipolar transistors or implanted diodes, using a CMOS process [61, 126]. Figure 1.20 shows a circuit topology consisting of two transistors for this purpose.

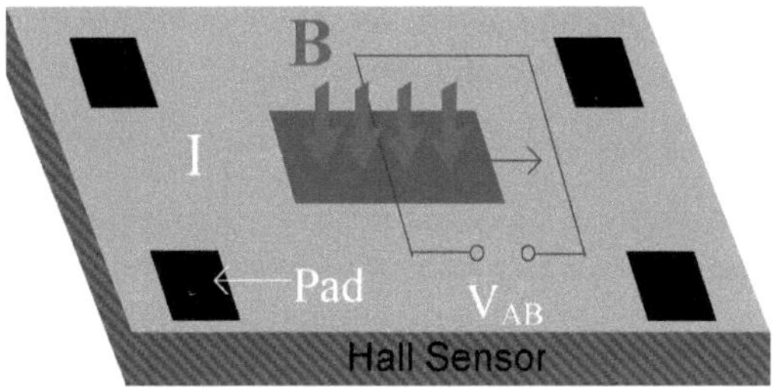

Fig. 1.18 Simplified diagram of an on-chip hall-effect sensor

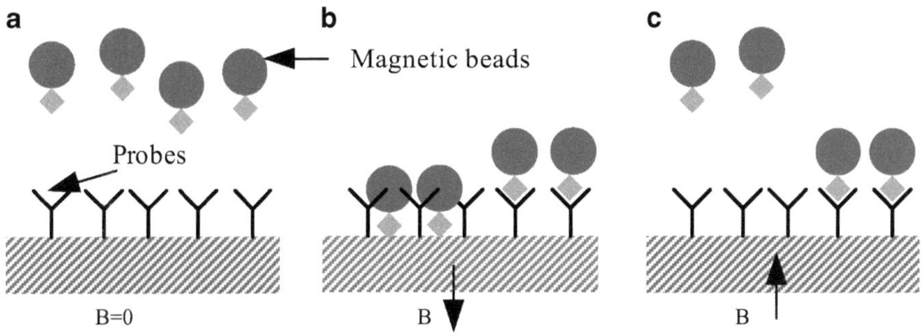

Fig. 1.19 CMOS Magnetic bead assay: beads are (**a**) injected, (**b**) pushed down and (**c**) non-bonded beads are removed

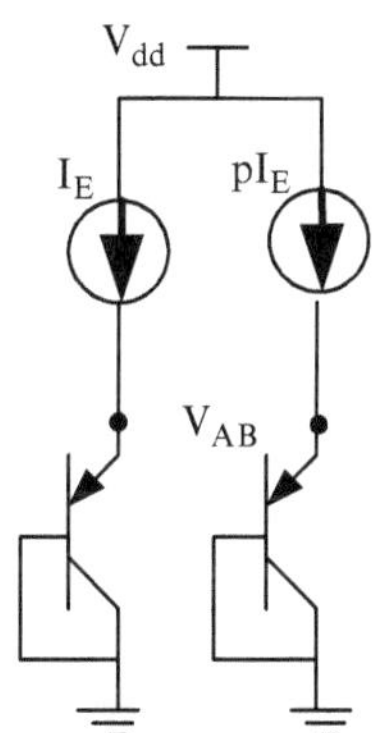

Fig. 1.20 A schematic of CMOS temperature measurement through parasitic PNP transistors

As shown in this figure, the difference between the base and the emitter voltage (V_{AB}) is obtained according to the following equation.

$$V_{AB} \approx \frac{KT}{q} Ln(p) \tag{1.2}$$

where K is Boltzeman's constant and q is electron charge. Also, it is assumed that the transistors are identical. In this simple topology, the ratios of injected currents remain constant and equal to p. As shown in Figs. 1.21, the substrate PNP transistor is formed in using an n-well CMOS process. This device can be employed to measure the temperature from −55°C to 125°C with an error of only 0.1°C, as reported in [127].

On the other hand, many efforts have been made to measure the temperature of the elements exposed to the analyte for calorimetric purposes, using either temperature sensitive beads resistors or thermoelectric methods [128]. The polysilicon layer in a CMOS process can be patterned for temperature measurement, but only as the embedded element. The thermometric method can be implemented through the topmost metal layer in order to directly measure the temperature of the analyte. The thermoelectric technique involves two different semiconductor or metal wires connected together in a loop, to form two heterojunctions. If there exists a temperature difference between the junctions, then a voltage will appear across them. By using aluminum and polysilicon as two conductors, a high precision thermoelectric sensor can be realized using CMOS processes [129]. Figure 1.22 shows a simple thermoelectric sensor coated with a polymeric sensing element.

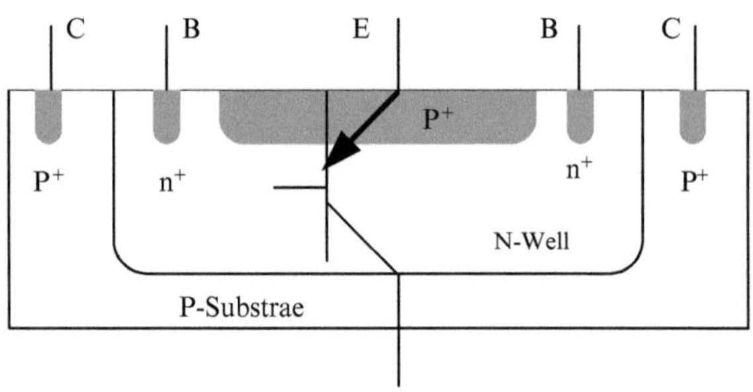

Fig. 1.21 Substrate BJT transistor in CMOS

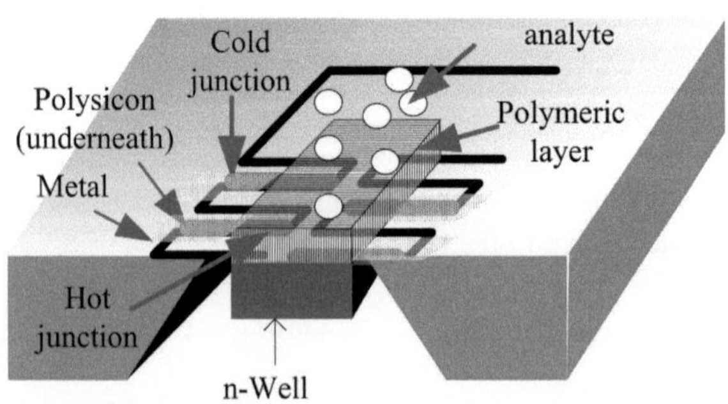

Fig. 1.22 Illustration of CMOS-based calorimetric sensor

As can be seen in this figure, the structure consists of two thermocouples, for temperature sensing, and a polysilicon hotplate to increase the temperature. This hotplate can also be implemented through a CMOS polysilicon layer. In order to thermally isolate this hot plate from other circuitries, the structure was generated using a back etching technique, where the silicon oxide was left as a membrane. This method can also be used to create a hotplate in order to control the temperature of a microenvironment at high temperature values [130, 131]. A schematic of micro-hotplate is shown in Fig. 1.23 [130] where four bridges are used to suspend the central hotplate. Of course, for many applications, such as cell culture, the temperature usually doesn't exceed 70°C and therefore there is no need to isolate the hotplate using laborious micromachining processes [132]. Such a micro-incubator was developed in John Hopkins University using standard CMOS and PDMS processes for biological applications [133].

All above mentioned CMOS sensors and actuators would require efficient microfluidic packaging to direct the fluid toward the sensing or actuating sites. Since the leakage of analytes (especially of charged molecules, as is the case with many bioanalytes) from microfluidic channels may increase the parasitic capacitances or resistances and thus affect the circuit characteristics, the implementation of microfluidic channels with hermetic bonding is very important. Additionally, such a microfluidic packaging should protect the bonding wires and pads from direct contact with bio-fluids. We will further explore this issue in Chapter 5.

1.3.7 Capacitive Sensing LoC

To date, several capacitive readout techniques, of various complexity, have been reported for autonomous MEMS-based capacitive sensor (MBCS) systems

Fig. 1.23 An illustration of on-chip hotplate

(e.g. accelerometer [134–137]) but there is a little published literature on the custom design of an on-chip capacitive sensor for LoC applications. The focus of this book is to describe the viability of on-chip capacitive sensor for LoC applications such DNA detection [64, 138] antibody-antigen recognition [70], cell monitoring [74], organic solvent detection [66], bacteria growth monitoring and ultrathin polyelectrolyte layer detection and detection of protein conformation, toxic chemical gases [139].

1.4 Objectives and Organization of Book

This unique multidisciplinary book provides a broad overview of the emerging field of CMOS-based capacitive sensing Laboratory-on-chip (LoC) technology by drawing in a range of disciplines: microelectronics, microfluidics and classical biochemistry [140]. The proposed book provides students and other researchers from academia and industry with knowledge of the design and implementation of hybrid microfluidic/CMOS capacitive sensors for several biological applications. In fact, this book, in addition to providing a review of recent literature, offers electrical engineers a comprehensive tutorial on the different aspects of fully electronic LoC systems, including bio-interfaces. To date, several books have described different LoC technologies, but less attention has been paid on fully integrated LoC systems. The emphasis of our proposed book is placed on full design and implementation of capacitive LoC systems using standard CMOS technology. The remainder of this book has been organised as follows:

- Chapter 2 will focus on the realization of sensing electrodes on CMOS chips. The required post-processing procedures for different applications are also put forward in this chapter. The electrical model of on-chip capacitive sensors and corresponding parasitic components are demonstrated.
- The biochemical procedures for the most important capacitive detection LoCs will be discussed in Chapter 3. This chapter introduces the concepts of bio-interfaces formed on electronic devices. The basic biochemistry and electrical properties of biological material, along with the corresponding biological procedures used to create bio-recognition layers, are described. We also discuss the most relevant bioassays for capacitive sensing purposes, including antibody-antigen interaction, DNA hybridization and cell viability, growth and settling mechanisms.
- In Chapter 4, readers would build on content from previous chapters to design accurate hybrid systems for various applications. In fact, the design strategy of capacitive sensors for LoC applications differs from other MEMS based capacitive sensors. Four different methods, including core-CBCM (charge based capacitance measurement), are described and their main practical issues, along with the advantages and difficulties of each method, will be discussed.
- Microfluidic packaging of CMOS chips is the focus of Chapter 5. This chapter concerns the recent progress made on the microfluidic packaging of CMOS

chips for bio-chemical applications. The microfluidic packaging is required to direct the fluid toward sensing sites and protect the die from having direct contact with the liquid. Additionally, direct-write assembly is described and its applicability for different applications is discussed.

- Present and Future of capacitive sensing technology for biochemical applications will be presented in Chapter 6. Afterwards, along with the promising future products of this emerging technology, the present research challenges of CMOS-based capacitive Lab-on-Chip are explained.

Chapter 2
Capacitive Sensing Electrodes

The capacitive sensing electrodes on the top of a CMOS chip serve as an interface between the microelectronic readout system and the biological/chemical analyte. These electrodes are directly exposed to the analyte or an intermediate layer which will be described in Chapter 3 (Fig. 2.1). The sensing electrode can be realized by a standard CMOS process. However for some applications, further micromachining procedure may be necessary.

In this chapter, we first describe the various configurations of sensing electrodes created above the CMOS chip for various applications, and then we will discuss the electrical model and associated parasitic capacitances of sensing electrodes.

2.1 On-Chip Microelectrode Configurations

The design and implementation of sensing electrode can be done using Virtuoso layout editor software using the top most metal layer (e.g. metal 6 in 0.18 CMOS process) [141]. Sensing can be performed using various configurations as shown in Fig. 2.2.

2.1.1 Passivated Electrodes

In standard CMOS technology such as 0.18 μm process (Fig. 2.2a), the passivation layers including silicon oxide, silicon nitride and polyamide are stacked in the last step of the process. A thin passivation layer (Fig. 2.2b) with uniform thickness is a good candidate for cell growth monitoring [142–144]. In this case, a large parasitic capacitance C_P (e.g. 0.5 μm CMOS process, 0.05 $fF/\mu m^2$ [75]) is created across the passivation layer. Therefore, the equivalent capacitance of the parasitic capacitance in series with a sensing capacitance C_S is approximately equal to sensing capacitance alone (see Fig. 2.2b).

E. Ghafar-Zadeh and M. Sawan, *CMOS Capacitive Sensors for Lab-on-Chip Applications: A Multidisciplinary Approach,* Analog Circuits and Signal Processing,
DOI 10.1007/978-90-481-3727-5_2, © Springer Science+Business Media B.V. 2010

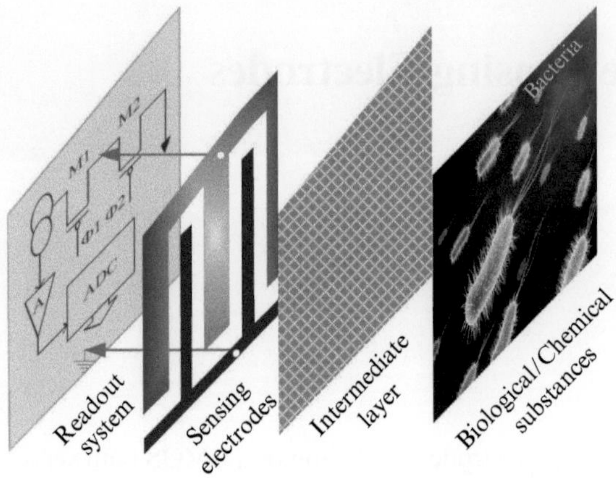

Fig. 2.1 CMOS based capacitive sensing LoC

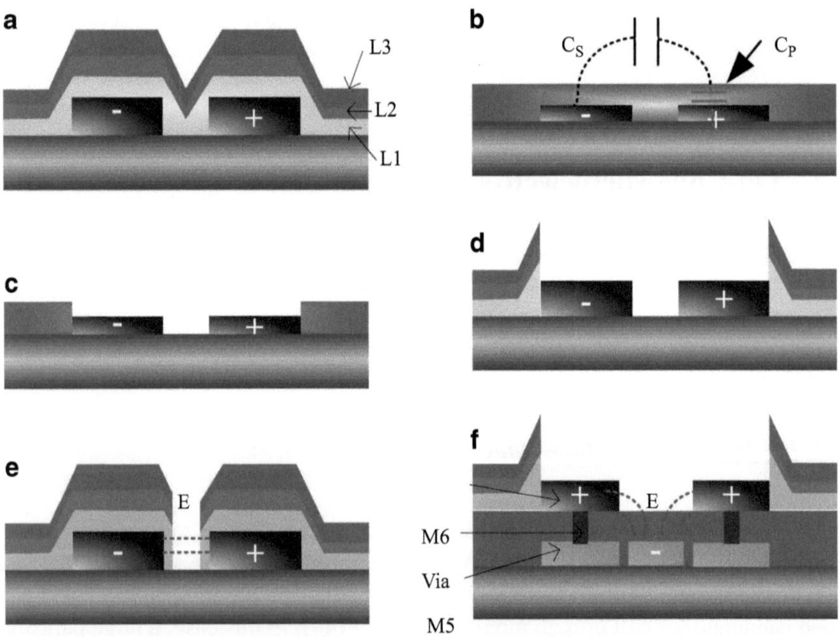

Fig. 2.2 Sensing electrodes realized atop a CMOS chip: (**a**) thick metal with three passivated layers, (**b**) thin metal with a passivated layer, (**c**) open-top thin metal electrode, (**d**) open-top thick metal electrode, (**e**) passivation electrode with a window in between the fingers, and (**f**) open-top electrodes with two metal layers

2.1.2 Unpassivated Electrodes

Aluminum is still the major material used for electrical contacts and interconnections in CMOS circuits. The top most metal layer in standard CMOS processes is basically made of aluminum plus a small impurity concentration of silicon (e.g. 0.18 CMOS, Al/1%Si, Al with 1% Silicon) [64, 145]. Aluminum is not widely used for biosensing purposes as opposed to gold and platinum. Due to the oxidizing property of the biological and chemical analytes, if a durable conductive electrode is required, only noble metals like gold (Au) and platinum (Pt) can be used but not aluminum (Al). Despite this, a native Al_2O_3 layer (≈10 nm) on the surface of Al is considered an advantage for biosensing applications [146]. This insulation layer makes the sensor more durable in typical biosensor environments. The viability of Al (along with the Al_2O_3 layer) as a sensing electrode has already been demonstrated for DNA detection and bacteria sensing [147, 148].

The passivation layers can be removed from the top of the electrodes in order to make a direct contact with the Al electrode (Figs. 2.2c). The "Pad" mask layer in this technology can be selected by the designer if needed. By increasing the height of the sensing electrode, the distribution of electric field and the performance of the biosensor can be improved significantly. As shown in Fig. 2.2d, by selecting a thick topmost metal layer in CMOS process and also by selecting a pad-etch mask, a larger space in between the electrodes can be made. A strong electrical field is created in this space because the electric field is generated by the parallel electrodes. The sensing electrodes realized in a thin topmost metal layer (Fig. 2.2c) can only detect the bioparticles in the top of the sensing electrodes, of course with less electric field lines.

2.1.3 Sensitivity-Enhanced Passivated Electrodes

The passivation layer in between the fingers can be removed in order to increase the sensitivity and dynamic range of sensing electrodes. As shown in Fig. 2.2e, the electric field in between the passivated electrodes become very strong, so that, the presence of bioparticles can significantly vary the sensing capacitance. The viability of this technique has already been demonstrated for the detection of liquid phase organic solvents [140].

2.1.4 Quasi Interdigitated Electrodes

In addition to selecting thick or thin metal layers, combining the electrodes from the two top most metal layers can improve the electric field (E) and subsequently the sensitivity. As shown in Fig. 2.2f, by selecting the topmost metal layer (e.g. metal 6 in 0.18 CMOS process) as the working electrode (including the sensing area), and by connecting ground to another electrode realized in the second metal layer (e.g. metal 5 in 0.18 CMOS process), a strong electric field can be created. This quasi interdigitated electrode with further polymer formation processes has already been reported for a CMOS-based gas sensor [149].

2.1.5 Gold Electrodes on CMOS Chip

Gold is used extensively as sensing electrodes for biomedical applications and is generally considered to be biocompatible [150–152]. While other less noble metal, such as platinum or iridium are oxidized to a depth of hundreds of nanometers, gold is a stable noble metal and a far better electron conductor than aluminum, copper or even silver. This highly defined and conductive surface of gold may be ideal for several biosensing applications including bacterial growth monitoring, virus detection, and DNA detection [153–155]. Furthermore, gold as a result of its unique surface chemistry allows for the self-assembly of organic molecules, through sulphur atoms. Such a self-assembled monlayer (SAM) can be used as a linker between gold sensing electrodes recognition layer [156, 157].

The gold layer can readily be fabricated using commercially available lithographic technologies on chip using CMOS compatible micromachining procedures at low temperature [158–162]. Figure 2.3a, b shows an illustration of gold electrode created above a CMOS chip. Section 2.3 presents the post-processing of gold on CMOS chip.

2.1.6 Microfluidic Channel Integrated Atop Sensing Electrodes

The sensing electrodes are usually realized on the same chip of capacitive interface circuit and a microfuidic channel is used to direct the biological fluid toward sensing site as seen in Fig. 2.4a [163, 164]. However, some rapid prototyping

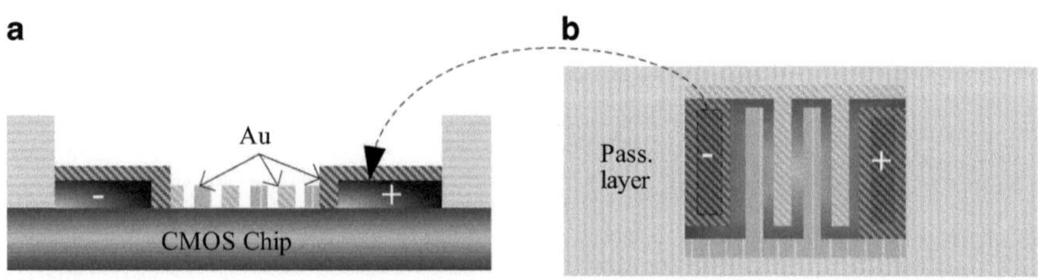

Fig. 2.3 Illustration of on-chip gold electrodes: (**a**) cross section and (**b**) top view of gold electrode on CMOS chip

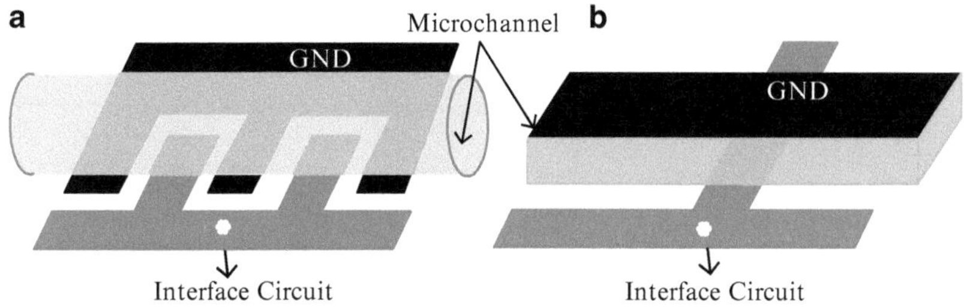

Fig. 2.4 Schematic representation of sensing electrodes incorporated with a microchannel: (**a**) interdigitated electrode and (**b**) configuration presented in [165]

methods have alternatively been reported to detect the bioparticles through the capacitive sensors created in between one electrode on the chip and another electrode between the chip and a grounded electrode above the chip as shown in Fig. 2.4b [165]. This figure shows a capacitive sensor including a common ground electrode on the top of microfluidic packaging. This technique can improve the dynamic range of the sensor by increasing the applied voltage.

2.2 Micromachining Gold Electrode on CMOS Chip

Extra micromachining procedures should also be performed to connect the deposited gold to underneath circuitry deep in the CMOS chip. Figure 2.5a–o depicts the micromachining steps to make gold electrodes atop a CMOS chip [158–162].

Step 1: Figure 2.5a shows a fabricated CMOS chip with two electrodes realized in two successive metal layers (e.g. Metal 5 and Metal 6 in 0.18 CMOS process). The pad-etch technique can be applied to remove the passivation layer.

After removing the passivation layers, the same passivation layers are deposited in order to create a small precise opening above the top most metal layer.

Step 2: A thin layer of silicon oxide is deposited (Fig. 2.5b).

Step 3: Silicon nitride is deposited (Fig. 2.5c).

Step 4: An opening above an Al electrode is created (Fig. 2.5d).

Step 5: Thereafter, the top of the passivation layer and the inside of the opening is coated in subsequently deposition with titanium (Ti) and titanium nitride (TiN), generally using a reactive magnetron sputtering technique (Fig. 2.5e). TiN is an excellent diffusion barrier material that is used in the advanced metallization step of integrated circuit manufacturing processes in particular for contact and via. TiN separates tungsten which is used as conductive via from titanium and silicon dioxide in order to avoid the interaction between contiguous layers.

Step 6: The opening is thereafter filled with a tungsten material (Fig. 2.5f).

Step 7: An etching process is performed in order to form a fine via between the Al_6 and the gold electrode (Fig. 2.5g).

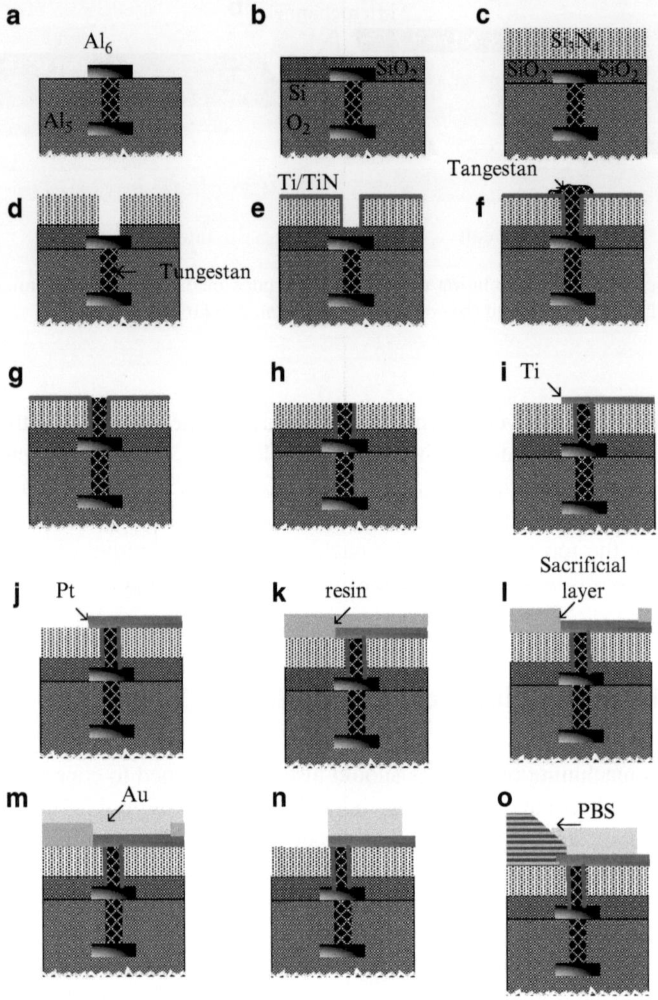

Fig. 2.5 On-chip gold deposition method: (**a**) fabricated CMOS chip, (**b**) silicon oxide and (**c**) silicon nitride depositions, (**d**) opening, (**e**) Ti/TiN deposition, (**f**) tungsten filling, (**g**) etching process, (**h**) Ti/TiN removal, (**i**)Ti deposition, (**j**) Pt deposition, (**k**) resin deposition, (**l**) sacrificial layer patterning, (**m**) gold deposition, (**n**) lift-off, and (**o**) PBC creation

Step 8: In the next step, the barrier layer is removed from the top of the silicon nitride passivation layer (Fig. 2.4h).

Step 9: Ti is deposited (Fig. 2.5i).

Step 10: The double layer of Ti/Pt is created above the tungsten via by depositing Pt above Ti (Fig. 2.5j).

Step 11: A thin layer of resin is deposited (Fig. 2.5k).

Step 12: This resin layer is patterned above Ti/Pt double layer(Fig. 2.5l).

Step 13: An appropriate thicknesses of Au is deposited (e.g. Ti: 50 nm/Pt: 50 nm/Au: 500 nm [22]) as shown in Fig. 2.5m.

Step 14: Using a lift-off process, the sacrificial layer along with Ti/Pt/Au on top of sacrificial layer are etched in order to create a gold electrode connected to the Al_6 through a tungsten via and Ti/Pt (Fig. 2.5n).

Step 15: A resist layer such as polybenzoxazole (PBS) is coated on the top of gold electrode and other parts of chip. This layer is thereafter developed and baked at a given temperature (Fig. 2.5o).

2.3 Electrical Model of Sensing Electrodes

A capacitive sensor with an interdigitated electrode on the top of a CMOS chip is implemented in order to detect the minute variations of capacitance on the sensing electrodes which are associated with large parasitic capacitances.

Figure 2.6 shows a SEM image of a sensing electrode where the passivation layers in between the electrodes have been removed [166]. The presence of analyte in between the fingers or above the electrodes can be detected by sensing electrodes. This figure shows the electrode from top and a 45° view. The width and the space between the fingers are about 10 μm approximately.

These parasitic capacitances and the equivalent circuit model of the sample are shown in Fig. 2.7. As seen in this figure, the capacitive and resistive properties of analyte are modeled by a parallel capacitance (Cs) and resistance (R). The parasitic capacitances across the passivation layer (C_{P1}), in-between the electrodes (C_{P2}) and in between silicon substrate and electrodes (C_{P2}) are also shown in this model.

In fact, R and C_s can be obtained from the parallel combination of a large number of small elements (δC_s, δR). As seen in Fig. 2.7a, b, by assuming the same values for δC and δR in each branch, the dropped voltage *V* on this parallel combination

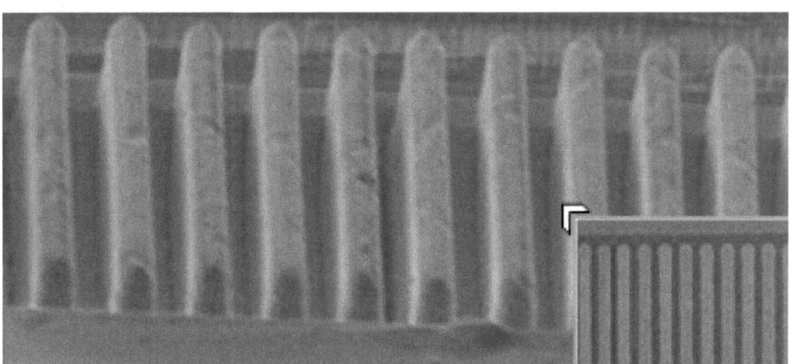

Fig. 2.6 SEM image of implemented sensing electrode on CMOS chip seen from two different view angles

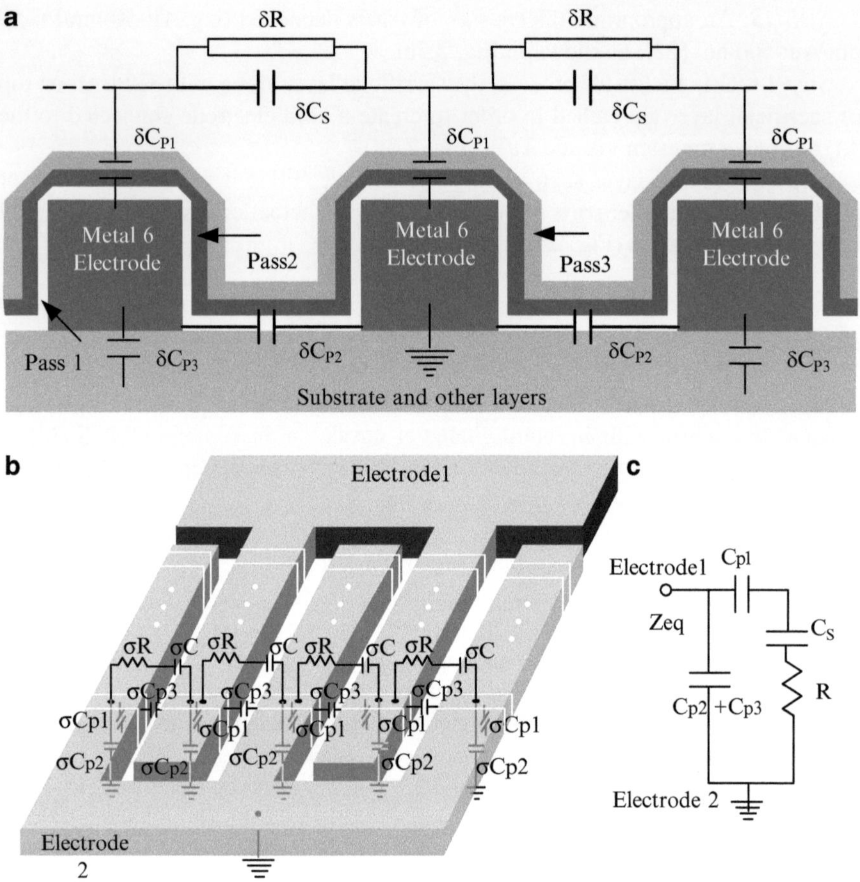

Fig. 2.7 Illustration of (**a**) and (**b**) parasitics generated on top of a CMOS chip (δC and δR are the partial parasitic capacitance and resistance respectively), and (**c**) its equivalent circuit

results in a current $I = I_1 + I_2 + + I_n = nI_1$. Considering the Laplace transform of $I_1(S) = V(S)/(\delta R + 1/\delta C_s S)$, therefore, $I(S)=nI_1(S) = V(S)/(\delta R/n + 1/n\delta C_s S)$.

In other words, by considering the above mentioned assumptions and $I_1(0) = I_2(0) = \ldots I_n(0) = 0$, the equivalent resistance and capacitance become $R \approx \delta R/n$ and $C_s \approx n\delta C_s$ respectively where the electrode is broken into n finite elements. Based on this discussion, the interdigitated electrodes with large number of fingers result in a large equivalent capacitance and a small resistance. The equivalent circuit shown in Fig. 2.7c is obtained for the capacitive electrode realized through a CMOS process. The parallel resistor and capacitor in this model is presented as an example. It should be mentioned that many other electrical models for biological and chemical samples have been documented for different applications [167].

2.4 Summary

In this chapter, we introduced with different methods used to realize sensing electrodes above CMOS chips. Corresponding to each biosensing method, the configuration of the sensing electrode is selected. The main advantage of these on-chip configurations is the simplicity of design and fabrication. However there are several limitations in selecting the materials exposed to analyte or the minimum dimensions. The modern nano-scale CMOS processes offer the implementation of large arrays of nanoscale electrodes; however in selecting the CMOS process, a compromise should be made between the minimum feature of the sensing electrodes, the applied voltage/current and the speed of sensor.

Chapter 3
Capacitive Bio-interfaces

In Chapter 2, we discussed the design and implementation of sensing electrodes atop CMOS chip. The sensing electrodes are incorporated with biological substances for sensing purposes as shown in Fig. 3.1.

As can be seen in this figure, an intermediate layer is coated above the sensing electrodes. The physiochemical changes of this layer for some biological assays can be detected by capacitive sensors. In some biosensing methods, the intermediate layer is formed by its own nature, for instance, due to the settling or growth of living cells on the surface of the sensing electrodes [168–170]. As a consequence of such a bioassay, a bio-film is created above the sensing electrodes. This biofilm creates a sensing capacitance where the positive and negative charges are separated by applied electric fields. The thickness modulation of this biofilm or the variation of its dielectric property due to the change of analyte concentration is detected by a capacitive sensor accordingly. In such biosensing methods, there is no selectivity, however in many other biosensing methods; a certain biofunctionalized intermediate layer is created for the detection of specific molecules or cells [171, 172]. Such an intermediate layer is called recognition element. In this approach, biological researchers are experiencing and exploring new bioassays which can be applied for the measurement of capacitance changes for biosensing purposes. On the other hand, the microelectronic engineering aspect of this research is focused on the implementation of capacitive sensors using standard CMOS technology.

Based on relevant literature, DNA hybridization detection, antibody–antigen binding recognition, on-chip living cell monitoring, bacteria growth monitoring, detection of protein-analyte conformation, polyelectrolyte layer detection and organic solvent monitoring can be performed using capacitive sensing methods [70, 173–181]. These biological methods are introduced in this chapter. While the discussion of the details of all these techniques is beyond the scope of this chapter, the focus here is on the design and implementation of protein based biosensors as an example. The main goal of this chapter is to give

E. Ghafar-Zadeh and M. Sawan, *CMOS Capacitive Sensors for Lab-on-Chip Applications: A Multidisciplinary Approach,* Analog Circuits and Signal Processing,
DOI 10.1007/978-90-481-3727-5_3,

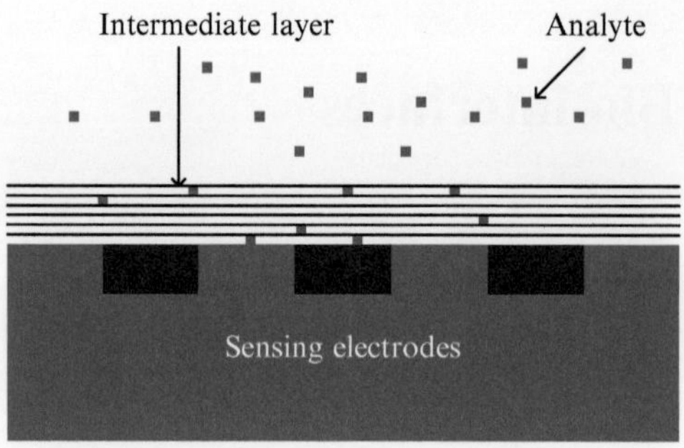

Fig. 3.1 Illustration of an integrated biosensor incorporated with intermediate layer

a clear idea to microelectronic engineers on how to incorporate the biochemical materials on CMOS chip and how a capacitance method can be exploited for such biological assays.

3.1 Biochemical Capacitive Sensing Methods

In this section seven different types of capacitive affinity biosensors are discussed. For each biosensing method, the biological/chemical principle, practical considerations and limiting factors of the CMOS processes are put forward.

3.1.1 Hybridization Detection

Genetic information is stored in DNA double-helical shaped molecules confined in cell nuclei [182]. A DNA code is a long and detailed message that instructs the cell how to make its vital proteins. Each code consists of sequences of four building blocks (A, G, C, T). Prior to reading out the DNA codes, several biological procedures should be accomplished to prepare the DNA sample from the cells as shown in Fig. 3.2.

These procedures include cell pre-filtering, fractionation, and focusing in order to separate rare cells (e.g. cancerous cells [183, 184]) from a blood sample and make them ready for bursting. Thereafter, the DNA strands are extracted from the cells. This cell lysis process is followed by other molecular procedures such as amplification, fragmentation and labelling [185, 186] to make the sample ready for DNA hybridization.

The hydrogen bonding of two complementary single strands of DNA such as *AGCATA* and *TCGTAT* is called hybridization. Thus, a double-stranded DNA or so-called renaturation process, occurs at a specific temperature and salinity conditions. In this process, the known strands are immobilized on a substrate (also called

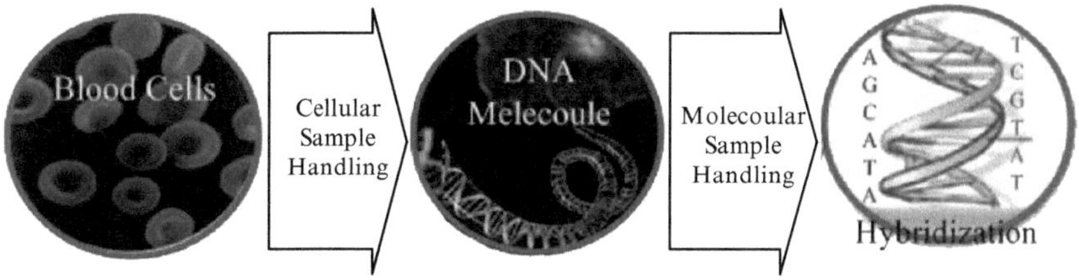

Fig. 3.2 Simplified diagram of biological procedure for DNA detection from blood samples

DNA probe) and the unknown strands wander up and down in the sample solution until stopping beside the according probes. The hybridization binding is specific hence hybridization serves as a high selective sequence detection mechanism (Fig. 3.3a). The presence of a double strand in the mixture can be detected optically by fluorescence tags which already labelled onto the input DNA molecules [90, 187–189].

A decrease in the electrochemical capacitance of an electrode–solution interface after being exposed to unknown fragments can also be detected by a low complexity capacitive sensor [190, 191]. Among the efforts to use capacitive sensor for DNA detection, Berggren et al. demonstrated a capacitance variation between 1 and 20 mF/cm^2 on functionalized gold electrodes with 26-base-long oligonucleotides, complementary to the targets [192]. Also Guiducci et al. presented a sensing method with two electrodes (Fig. 3.3b) instead of three electrodes in conventional electrochemistry (counter, reference and ground electrodes) [193]. Figure 3.3b shows the creation of double layer capacitance (C1, C2) in parallel with resistors (R_{P1}, R_{P2}) above both sensing electrodes. The conductivity of electrolyte is also modeled with a resistor R_S. As a follow up of this work, they reported the realization of an array of gold electrodes and subsequent readout circuitry using a CMOS process [194]. For this, a low temperature sputtering method was used to realize gold electrodes on CMOS chip. It should be mentioned; that there are some efforts to immobilize DNA on other CMOS compatible material such as silicon dioxide [195] or aluminum [64]. However, the direct immobilization of DNA using standard CMOS without extra post-processing is an important challenging issue to the best of our knowledge.

3.1.2 *Antibody–Antigen Recognition*

Antibodies are proteins deployed by the immune system to identify and neutralize foreign bodies, such as viruses [196, 197]. A small region at the tip of each antibody is extremely variable, allowing millions of antibodies with slightly different tip structures to exist. Each of these variants can bind to different target, known as an antigen. When an antigen binds to immobilized antibody layer such as anti-human

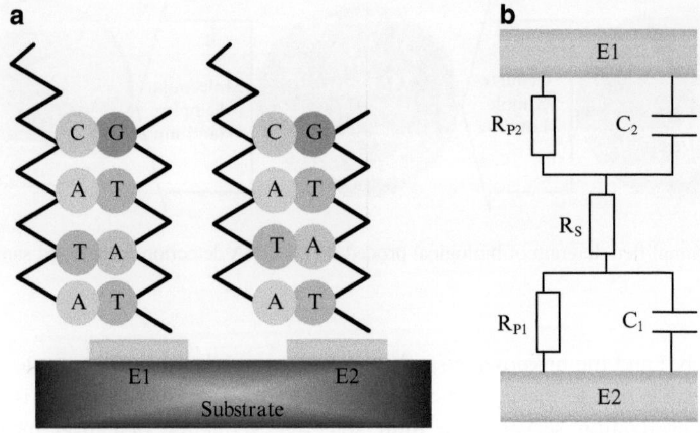

Fig. 3.3 DNA hybridization: (**a**) illustration of on-chip DNA detection and (**b**) electrical model of sensing electrode exposed to DNA solution

transferring [198] at the substrate, a modulation occurs in thickness or dielectric constant parameters that can be detected through a capacitive sensor (Fig. 3.4a) [199–201]. Of course, surface plasmon resonance sensor (SPRS) is a well established optical method to detect the mass variation resulting from antibody–antigen interaction [202, 203]. However, the realization of a technically complicated SPRS using a standard CMOS process is not favorable. Moving toward a low complexity approach, Balasubramanian et al. used a capacitive method to detect viruses through a standard enzyme linked immunosorbent assay (ELISA) [70]. The antibodies were immobilized on silicon dioxide where thick metal layer in CMOS was used to increase the sensitivity of detection. Also, a temperature treatment was performed prior to sensing in order to decrease the ionic conductivity around the sensing sites. By applying the temperature treatment process, the electrical model of this sensor without any resistor can be shown in Fig. 3.4b. In this figure, C_{P1} and C_{P2} are the capacitors created across the passivation layer and C_S is the sensing capacitance.

3.1.3 *Living Cells Monitoring*

It has been recognized in bio-electrochemical studies, that the morphological states of biological cells have a strong correlation with their electrical properties and also the existing membrane potential can be considered as an indication of living cells [204]. Among very few works for on-chip cell monitoring, a CMOS capacitive sensor has been reported for a cell based on above mentioned bio-electrochemical observations. As seen in Fig. 3.5a, the dielectric property [204] or

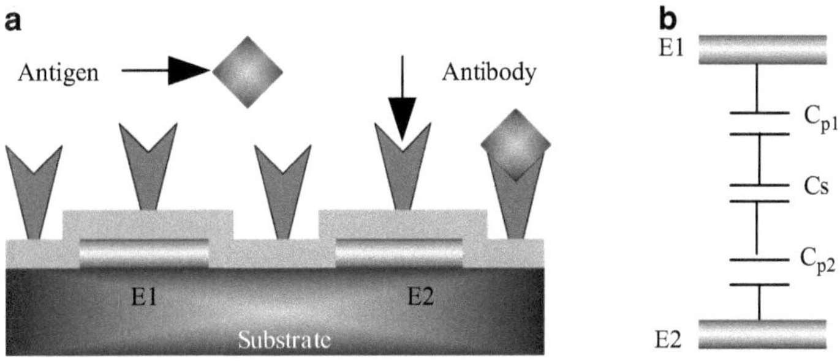

Fig. 3.4 Antibody–antigen binding recognition: (**a**) illustration of on chip antigen–antibody recognition and (**b**) equivalent circuit

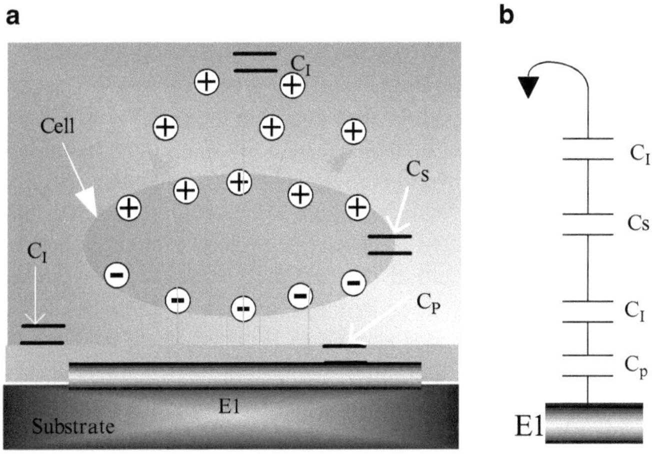

Fig. 3.5 Living cell monitoring: (**a**) illustration and (**b**) equivalent electrical model

surface charge [75] of a living cell can play essential roles in this measurement technique. These charges are moved upon exposure to an electrical field resulting in a dipole moment. The contact of the cells with the passivation layers on top of the sensing electrodes and the created dipole moment are significantly detected by a capacitance change (e.g., 1 fF capacitance variation on 400 μm^2 electrode [75]). In the model shown in Fig. 3.5b, the sensing capacitance, interfacial capacitance and parasitic capacitance across the passivation layer are denoted C_S, C_I and C_P respectively. As will be described in Chapter 4, the corresponding capacitance with living cells are detected using a floating capacitor technique. In another effort, another method has been reported for cell localization using an array of parallel electrodes – one common electrode covering the microfluidic structure, and an array of addressable electrodes on top of the CMOS chip [74]. In this technique the dielectric change in proximity of the sensing electrodes in the presence and absence of living cells is measured.

3.1.4 Organic Solvent Sensors

The detection of organic solvents is critical in the monitoring of food, drug, soil and water samples [205], owing to the toxicity of such chemicals. The dielectric constants of solvents can be used as the selectivity and/or sensitivity factor for their detection purposes. On the other hand, a low electrical conductivity is expected for most polar solvents. These properties can be considered as advantages to exploit a capacitive sensor (not impedometric sensor) to monitor the organic solvents. In addition, as the sensing electrode are covered by passivation layers, such an insulation of the sensing electrodes drastically decrease the leakage current (due to low conductivity of solvents) and consequently improve the capacitive sensing property. A liquid phase solvent can be injected into a microfluidic channel above sensing electrode as shown in Fig. 3.6a. The electrical model of the sensing electrodes exposed to the organic solvents is shown in Fig. 3.6b. This figure shows the sensing capacitance (C_S), the interfacial capacitance (C_I) created on the surface of the passivation layer and the parasitic capacitance (C_P) created across the passivation layer. The large value of C_I and C_P can be discounted in the series with C_S.

The sensing electrodes can be realized by a thick top most metal layer in order to create a strong electric field and consequently high sensitive capacitive sensor in between the fingers. In this direction, Ghafar-Zadeh et al. reported a capacitive sensor that include an interdigitated electrode for organic solvent detection [206].

It is obvious once the solution reached the electrodes, that the variations of sensing capacitance could be monitored as a function of thickness (H_t) of the solution above the sensing electrodes as long as $H_t \ll w$, where w is the width of the fingers. In fact, by injecting a solution with higher H_t the capacitive sensor was saturated as shown in this Fig. 3.7a. In this simulation, a solution with certain dielectric constant value is injected in a microfluidic channel similar to the one fabricated by Direct-Write Fabrication Process (DWFP) [207, 208] as shown in Fig. 3.7b, c. Based on this simulation results, the sensor can detect the bioparticles up to a distance ≤w. For gaseous solvents, a polymer sensing layer is used on top of electrodes as an interface between the electrodes and the solvent. Therefore, the density of diffused molecules and consequently the dielectric constants of the

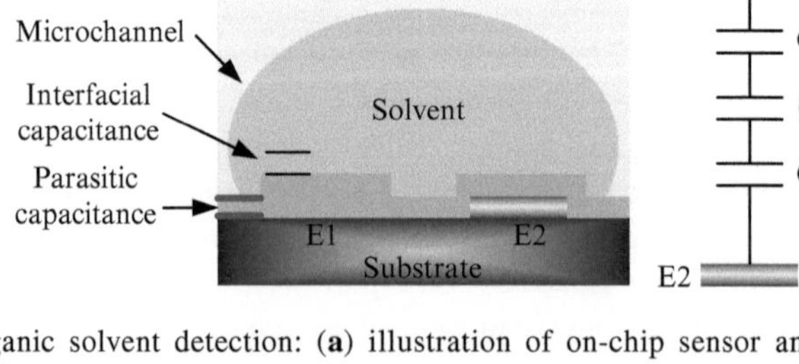

Fig. 3.6 Organic solvent detection: (**a**) illustration of on-chip sensor and (**b**) equivalent electrical model

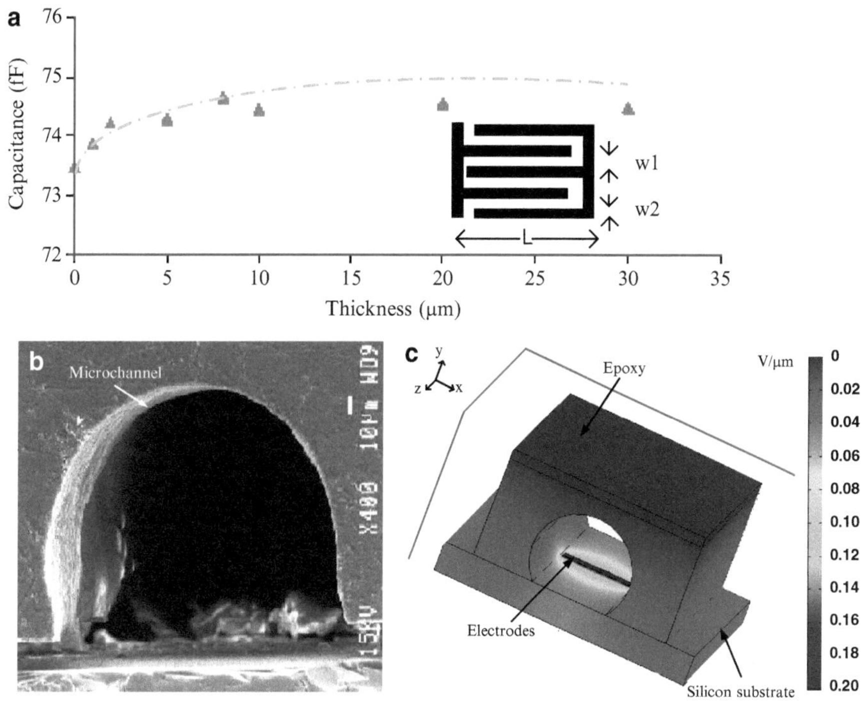

Fig. 3.7 Simulation and modeling results for interdigitated capacitive electrodes (**a**) simulation results of extracted capacitance in between the electrodes for different thickness value of the thin layer of liquid in the channel, (**b**) SEM image of a microchannel fabricated using direct write process, and (**c**) the FEMLAB model of microchannel

insulation layers above the electrodes will proportionally vary with respect to the concentration of solvents [73].

Hagleitner et al. proposed a CMOS capacitive sensor for the detection of toluene and ethanol using polyetherurethane (PEUT) as the polymeric sensing layer [72]. As shown in Fig. 3.8a, a quasi interdigitated electrode was incorporated with PEUT. Another metal layer was used as a stop layer for pad-etching of passivation layers in between the electrodes. The electrical model of this sensor is purely capacitive as depicted in Fig. 3.8b.

3.1.5 *Bacteria Growth Monitoring*

Growth of bacteria is the division of one bacterium into two bacteria. The growth rate depends on several parameters including environmental temperature and pH of the solution [209]. The growth and/or settling of bacteria create a bio-film atop the sensing electrode. Therefore, the growth of bacteria varies the dielectric constant and conductivity of this bio-film over time.

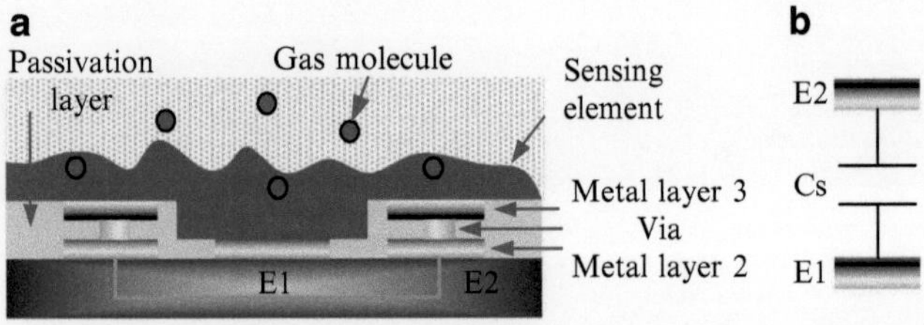

Fig. 3.8 Gaseous organic solvent detection: (**a**) illustration of an on-chip structure and (**b**) electrical model

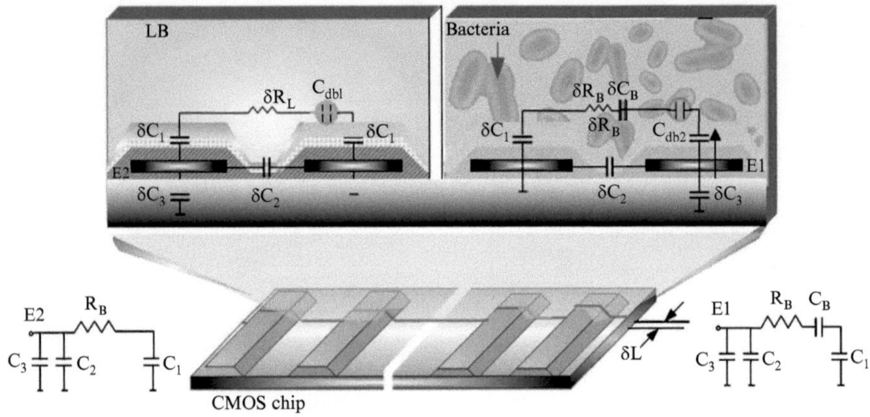

Fig. 3.9 Electrical model of LB medium and bacterial solution

A medium solution is used to grow bacteria. For example, Luria-Bertani (LB) is the medium of *Escherichia coli* (abbreviated as *E. coli*). As shown in Fig. 3.9, the bacteria and LB solutions can be directed toward the sensing sites atop the passivated electrodes using two microfluidic channels. Figure 3.9 additionally shows the electrical equivalent models of the sensing and reference electrodes. R_B and C_B are the corresponding capacitance and electrolyte resistance of the bacteria solution. In this simplified model, the effect of the double layer capacitances (C_{db1} and C_{db2}) is discounted in series with other small capacitances. In fact, the double layer capacitance, across a very thin layer of biofilm result in a large capacitance. The parasitic capacitances C_1, C_2 and C_3 are also associated with the sensing electrode and substrate. In this figure, the prefix δ shows that the resistances and capacitances values belong to a thin slice (δL) of the sensing electrode.

C_B varies over time due to the settling and growth of bacteria where it is assumed that other parameters including R_L, R_B and parasitic capacitances are constant. Therefore, Bacteria Growth Monitoring (BGM) can be performed by measuring C_B. On the other hand, the LB solution can be modeled with a resistance R_L as shown in Fig. 3.9. It should be mentioned that R_B is also varied due to variation of electrolyte volume with respect to the whole bacteria solution. However, this variation does not affect the measurement of C_B by the charge based capacitive sensor [210, 211].

3.1.6 Polyelectrolyte Monolayer

The ultra thin multilayers of charged molecules are simply fabricated by alternatively injecting (or dipping etc.) the dilute aqueous solutions of positively and negatively charged molecules or so-called polyelectrolyte onto the substrate with a rinsing process in between [212]. By repeating this procedure, an arbitrary number of times, the ultrathin polyelectrolyte layers (PLs) can be built up due to the electrostatic attraction between them. This process has already been reported for different applications such as glucose detection [213, 214] and drug delivery [215] by controlling antifouling property. Based on literature, a significant dielectric property is recognized in between the layers with opposite charges [216]. The thickness of ultrathin PLs is of the order of a few nanometers [212]. Negatively charged DNA molecules or positively charged Chitosan are widely used as biological polyelectrolyte for biosensing purposes [217]. This low temperature process with no destructive chemical solutions is also CMOS compatible, as demonstrated by Ghafar-Zadeh et al. [218] using polycation chitosan and polyanion alginate polyelectrolytes. This property can be exploited for capacitive detection of ultrathin layers.

It is worth to mentioning that the charge layer covering the particle can traditionally be detected by the Zeta potential measurement (ZPM) method [219]. As the principle of ZPM method is established on the fact that the charge of moving particles include an electric field which can be determined by measuring their speed and direction, this method is not applicable for the charged layers on the surface of a substrate or CMOS chips. This issue can be addressed by using a capacitive technique to detect the charged layer and/or dielectric double layers formed on a chip. First, the detection of these polyelectrolyte based layers using the simple capacitive technique has already been investigated using a microfabricated electrodes [181]. Thereafter an integrated capacitive sensor system was employed to show the functionality of this method for CMOS sensors [220].

Figure 3.10 shows a simple method of creating ultra thin multilayer of charged molecules on CMOS chip. The positively and negatively charged polyelectrolytes are alternatively staked. The rinsing process in between the depositions is definitely important in order to remove non-matched molecules and consequently create a very uniform and thin layer of molecules. Also this figure shows a rinsing process in between. As demonstrated in this work, up to five layers could be stacked up. The first layer is a positively charged polyethyleneimine (PEI) layer which is formed on the surface of the chip to initiate the sequential adsorption of the weak polyelectrolytes. This layer is followed by aliginate, chitosan, alginate and chitosan layers subsequently.

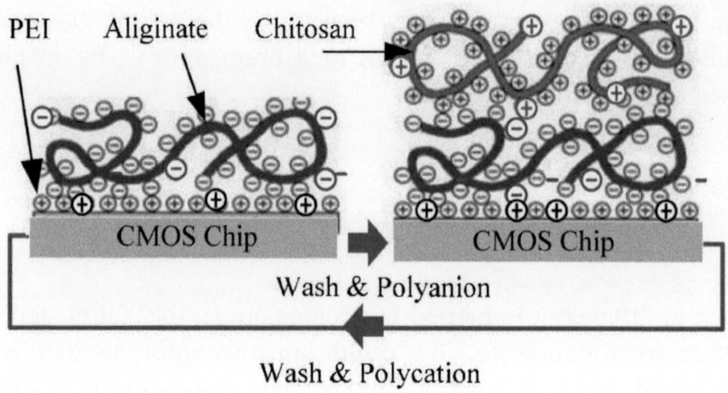

Fig. 3.10 Schematic of the formation of PLs on CMOS chip

Figure 3.11a, b also show the electrical model of sensing electrodes exposed to polyelectrolyte layers.

As shown in these figures, C_{p1} and C_{p2} represent of the parasitic capacitances between the electrodes and analyte and C_{EG} and C_{ES} are the parasitic capacitances between the electrodes and the substrate. Based on this model, the formation of the first PL results in a large capacitance $C_S = C_{PL1}$ (see Fig. 3.3a). In this case, the total capacitance C_{M1} is approximately equal to $(Cp1 \| Cp2 + C_{EG} + C_{ES})$. On the other hand, the presence of the second PL of opposite charge creates a dielectric double layer allowing the electric field to pass through. This double layer results in $C_{M2} \approx C_{S0} \| Cp1 \| Cp2 + C_{EG} + C_{ES}$ where C_{S0} is the sensing capacitance created above the surface. It is obvious, due to $Cs \ll Cp1 \| Cp2$ and consequently $C_{PL2} < C_{PL1}$ that the formation of an extra layer can cause a significant change in C_{M1} and C_{M2} and therefore can be detected by a capacitive sensor significantly.

3.1.7 Detection of Protein Conformation

Proteins are good candidates for creating biosensors, as their biologically active macromolecules can be easily introduced to the living environment and their unique binding properties can be exploited to create high-specificity biosensor. There are many proteins which can be used for identifying various biomolecules [221],[224]. Binding of biomolecule with their corresponding proteins induces a physiochemical change which can be detected using electrical or optical techniques [222]. As shown in Fig. 3.12, the proteins should be immobilized on a substrate which are exposed to a solution including target molecules. The conformational change of proteins upon binding with corresponding molecules can be detected through a variety of solid-state transduction devices including piezoelectric devices, capacitors, cantilevers, and Ion ISFET [223, 224]. As reported by Park et al. [225], ISFET is a good candidate to measure the surface charge of protein layer exposed to target molecules. In that work, an ISFET device was implemented using a 0.5 μm CMOS

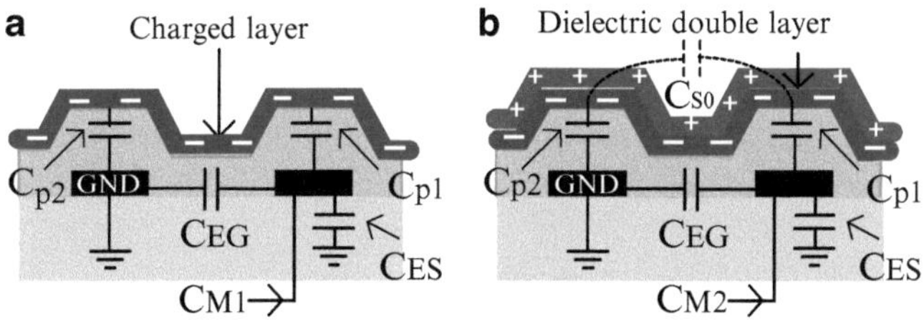

Fig. 3.11 Schematic of sensing electrodes along with associated parasitic capacitances on CMOS capacitive sensor

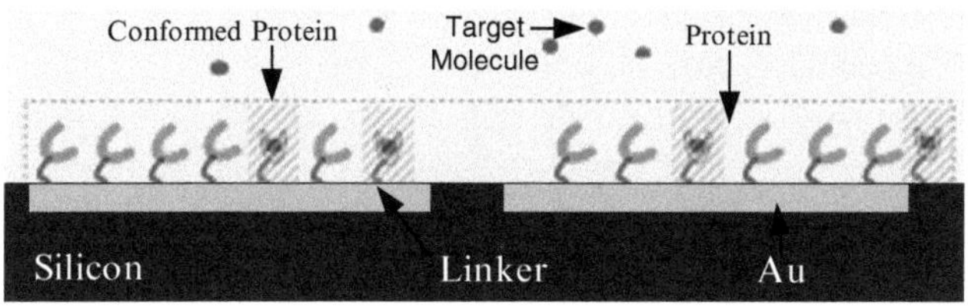

Fig. 3.12 Illustration of a protein based recognition element

process [226] incorporated with a maltose binding protein. The functionality of this technique was successfully demonstrated by showing a significant change in the drain current of the ISFET exposed to the maltose in Phosphate Buffered Saline (PBS) buffer. However, the pH sensitivity of ISFET devices may deteriorate the performance of this sensing system in monitoring different concentrations of target molecules. Impedometric or capacitive readout methods are other alternatives to detect the conformational change of proteins [227] which can be applied for many applications including glucose monitoring [228].

In the next section, we focus on a reagent less protein-based continuous glucose monitoring (CGM) by describing the biochemical procedures and experimental results. This section might be advantageous for engineering researchers who are interested in the design and implementation of biosensors but are lacking a background in biochemistry.

3.2 Design of Recognition Element: An Example for Continuous Glucose Monitoring

In this section, we describe the biochemical procedures used to create a recognition element above sensing electrodes. The discussion is exemplified by a protein based glucose sensor based on our recent experiences. This section presents the details of biochemical assays for very important biosensing applications.

3.2.1 Introduction to Glucokinase-Based Glucose Sensor

Blood glucose monitoring is very important for millions of diabetes sufferers. The development of an accurate CGM system is the key challenge in a closed-loop artificial pancreatic system to control insulin level for type 1 diabetes patients who experience loss of the insulin-producing beta cells in the pancreas [227]. To date several companies have successfully commercialized highly accurate glucose monitoring systems (e.g. FreeStyle Lite®, Abotte Inc. [228]) mostly using disposable electrodes incorporated with glucose oxidase enzyme that catalyzes the oxidation of glucose [229, 230]. As a consequence of this catalytic enzymatic reaction, the concentration of glucose is obtained by measuring the released charges. However, a CGM should be performed using a reagent-less method. Human glucokinase enzyme can be an ideal candidate for this purpose. Glucokinase (ATP: D-glucose 6-phosphotransferase) plays a pre-eminent role in the regulation of hepatic glucose metabolism and is considered as the glucose sensor of the insulin secreting pancreatic islet β-cells [231, 232]. This enzyme selectively binds the optical isomer of glucose but with a low affinity and a characteristic sigmoidal saturation curve which contributes to its role as a glucose sensor. Significant conformational change of glucokinase from super open to fully closed states upon glucose binding is reported in the literature (see Fig. 3.13a, b) [233].

Other research groups have studied the catalytic activation of human glucokinase by substrate binding-residue contact involved in the binding of D-glucose to the super-open form and conformational transitions [234]. This large conformational change of glucokinase as a result of substrate binding leads to a change in the overall dipole of the glucokinase enzyme. The conformational changes that occur in response to the binding event between glucokinase (GLK) and glucose can be measured using an impedometric technique [225].

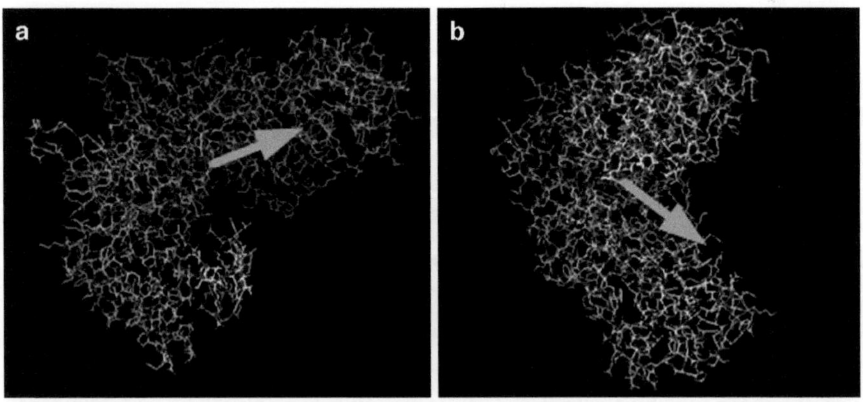

Fig. 3.13 Showing difference in net dipole (1060 vs 1023 DEBYE) and in direction of dipole (up vs down) between the (**a**) open and (**b**) closed form of human glucokinase enzyme

A wild-type and mutant glucokinase (GK) can covalently be coupled on screen printed carbon electrodes [226] or gold electrodes [235].

3.2.2 Immobilization of Glucokinase on Gold Electrode

Alkanethiol Self Assembly Monolayer (SAM) on gold is highly reproducible and well-characterized to generate well-defined organic surfaces with highly alterable chemical functionalities displayed at the exposed surface [236, 237]. SAMs have been widely used for studying biological and chemical processes, molecular interactions and, electrochemical and electronic applications [238]. The octane dithiol assembled monolayer on gold can be coupled with nitrilotriacetic acid Ni^{2+} linker (Fig. 3.14a) which shows a strong affinity for histidine-tagged recombinant human glucokinase enzyme (Fig. 3.14b).

Prior to the synthesis of the self assembled monolayer of 1,8-octanedithiol shown in Fig. 3.15, the surface of the gold coated electrode was washed with piranha solution (mixture of 66% concentrated sulphuric acid and 33% hydrogen peroxide) for 5 min. Then the surface was rinsed thoroughly with deionized water, absolute ethanol and dried under nitrogen gas. After every coupling, thorough washing was also carried out with deionized water and absolute ethanol to ensure complete removal of physisorbed material.

Maleimide-C3-NTA and $NiCl_2$ were added as shown in Fig. 3.15, prior to coupling with the His-tagged glucokinase enzyme. This procedure can be performed using the following steps.

- *Step 1*: The electrodes were cleaned using a mixture of H_2SO_4 and 33% H_2O_2 in a 3:1 ration and then the electrodes were rinsed thoroughly with de-ionized water followed by absolute ethanol. The electrodes were then dried under nitrogen gas.
- *Step 2*: Self assembled monolayers of 1,8-octanedithiol are created on Au electrodes by immersing the gold substrates in a 1 mM ethanol solution at ambient condition overnight.
- *Step 3*: SAM coated electrodes ere rinsed thoroughly with ethanol in order to remove any physio-absorbed material.
- *Step 4*: SAM coated electrodes are immersed in a 10 mM solution of maleimide-C3-NTA (pH 8.5 in Tris-HCl buffer) overnight, in order to perform the coupling reaction of maleimide-C3-NTA and a sulfhydryl surface.
- *Step 5*: Electrodes with NTA ligand are rinsed with deionized water and methanol in order to remove any physio-absorbed material.
- *Step 6*: Electrodes with NTA ligand are immersed in 200 mM $NiCl_2$ in order to load nickel cations (Ni^{++}).
- *Step 7*: Electrodes with Ni^{++} on the top of surface are rinsed with deionized water to remove the residual metal and ion solution.
- *Step 8*: His-tagged GLK proteins are coordinated to the Ni^{++} when the surface treated electrode in step 7 is immersed in 100 mM solution.
- *Step 9*: The created sensing electrodes were cleaned with deionized water and buffer.
- *Step 10*: The implemented protein-based sensor was freshly preserved in the buffer.

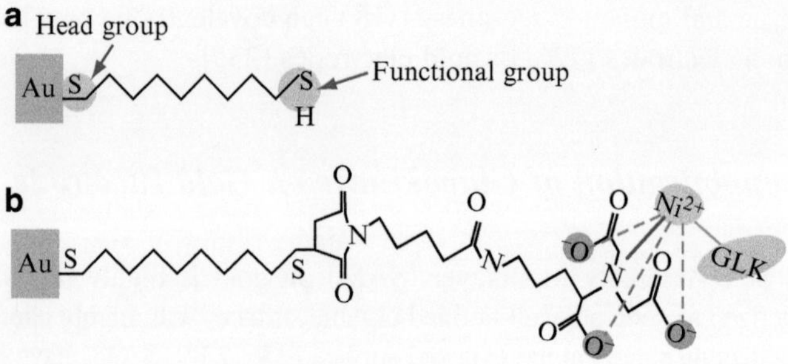

Fig. 3.14 Linker: (**a**) SAM of octanedithiol on Au (**b**) SAM of octanedithiol attached with melemide-NTA_Ni^{2+} linker which link His-tag glucokinase

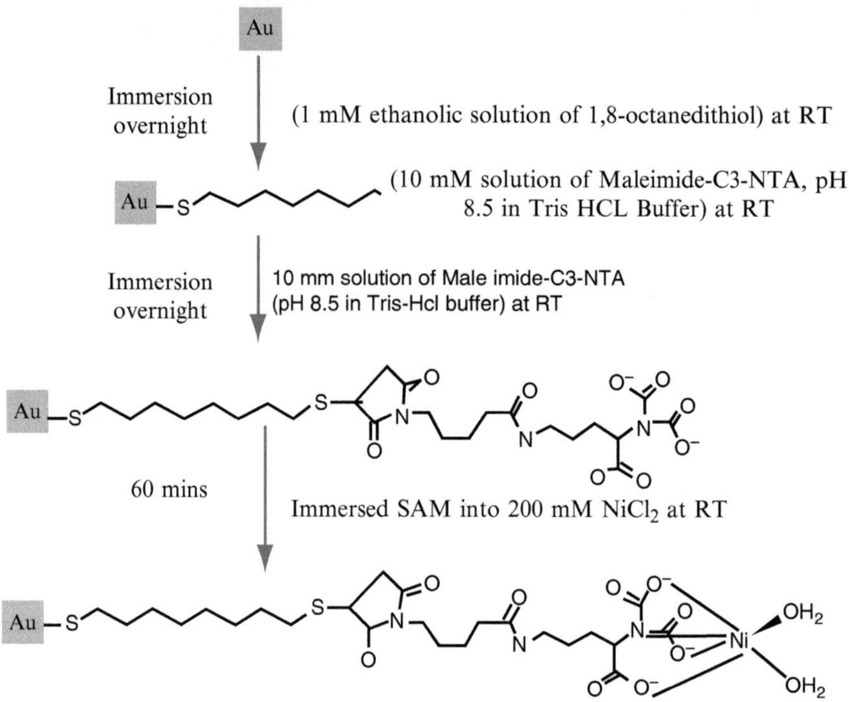

Fig. 3.15 Surface reaction scheme showing attachment of NTA functional groups on to the free sulfhydryl groups of 1,8 octanedithiol SAM on gold electrode

3.2.3 Glucose Testing

The linked proteins onto the gold electrodes were exposed to a solution including glucose molecules. In fact, the glucose-induced conformational change of the GLK protein induces a difference in its dielectric property which in turn could be detected

Fig. 3.16 SEM image of gold electrodes coated by protein based recognition element

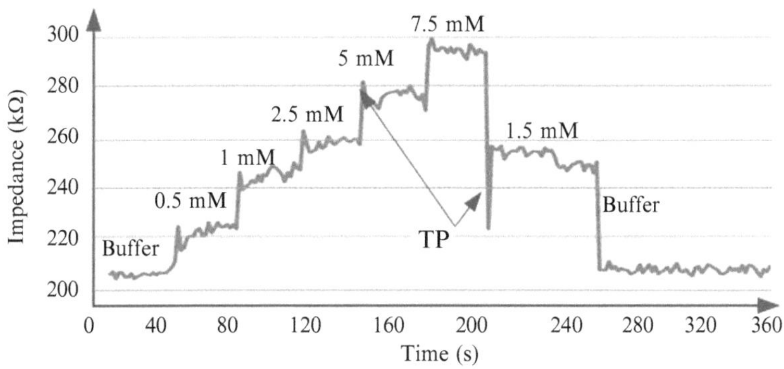

Fig. 3.17 Continuous-time measurement results for different concentrations of glucose

through an impedance sensing system. SEM images in Fig. 3.16 show the inter-digitized gold electrodes incorporated with the above mentioned GLK recognition layer.

The functionality of the system is demonstrated in the Fig. 3.17. This figure shows the impedance variations of the biosensing electrodes over time at $f \approx 30$ kHz for different concentrations (0.5, 1, 2.5, 5, and 7.5 mM) where the measurement electrode is rapidly removed from the testing tube and immersed in another testing tube. Based on these testing results, glucose can be sensed in the physiologically relevant range of 0.5 and 7.5mM and, also, a change in glucose concentration occurs in less than 10 s. In these experiments, the electrode were immersed in testing tubes and rapidly removed and inserted into another test solution. During this testing, the electrodes were not in the test solution for a short duration and therefore the impedance measurements shows ambiguous response in the highlighted regions and Transition Point (TP) as shown in Fig. 3.17.

The above mentioned method presents a low cost and portable reagent less biosensor for continuous glucose monitoring. The measurements were performed through a low-cost impedance based readout system using a commercially available impedance reader. The proposed reagent-less and replenishable technique with microelectronic integrated sensor is well positioned for further study towards the fabrication of implantable glucose sensors.

It should be mentioned that the main components of impedance which is varied as a function of glucose is the capacitance. This capacitance can be measured by a capacitive sensor. The capacitive measurement of such a protein conformations is a yet unmet challenge of particular interest.

3.3 Summary

Based on the discussions throughout this chapter, capacitive sensors have an enormous scope of applications in medicine and biotechnology by detecting thousand of types of viruses, bacteria or organic solvents as well as millions of DNA fragments. These high demand applications could be essential technology drivers for batch production and continued research for new diagnostic devices. Of course, for each application, it is required that a specific recognition element can be implemented.

Chapter 4
Capacitive Interface Circuits for LoC Applications

The design criteria of capacitive interface circuits for LoC applications differs from conventional MEMS based applications such as acceleration, vibration or pressure [238]. As already mentioned in Chapter 2, a movable sensing electrode should be implemented through MEMS procedures and then bonded to an interface circuit for measurement purposes, but the surface electrodes can be directly realized atop integrated circuit chip fabricated through standard CMOS technology [240]. The emphasis of this chapter is placed on describing the difference between a MEMS based capacitive sensor (MBCS), a Lab-on-Chip based capacitive sensor (LBCS) and on introducing various circuit design techniques for LBCSs.

4.1 LBCS Versus MBCS

In this section, the main design features and operational characteristics of LBCS are briefly described and compared with MBCS.

4.1.1 Instant Measurement

A MBCS is often operated over a continuous and lengthy measurement time. For example a sensory device fixed to a piece electro-machinery should be able to monitor the vibration for weeks or months without error. A built-in self-calibration module is often incorporated in interface capacitive circuit to correct the accumulating errors. A LBCS does not suffer from such a problem. Actually, in a LBCS, the subtraction of sensing capacitances in the presence rather than the absence of analyte is measured for only a short period of time. As seen in Fig. 4.1a, if a random error (Δ) appears on an MBCS, it should be cancelled. Otherwise, all capacitance (C_S) measurements will thereafter be affected. Figure 4.1b shows capacitance ($\acute{C}_S$) measurements on the same analyte at different times. As shown in this figure, a

E. Ghafar-Zadeh and M. Sawan, *CMOS Capacitive Sensors for Lab-on-Chip Applications: A Multidisciplinary Approach,* Analog Circuits and Signal Processing,
DOI 10.1007/978-90-481-3727-5_4,

time-variant random error Δ(t) can change the measurement results before and after applying a sample (by assuming Δ(before sample injection) = Δ(after sample injection)) but the subtraction of measured capacitances ΔĆs remains constant.

4.1.2 Aqueous Measurement

A MBCS is prevented from having direct contact with fluids, dust or even air by an appropriate vacuum chamber in order to obtain reliable measurement results, while a LBCS is always exposed to aqueous biological or chemical samples. For this, the presence of non-bonded bioparticles or other remnants in the channel may alter the design characteristics of the sensor based chip. For instance, as seen in Fig. 4.2, the presence of a small droplet creates a parasitic capacitance which is substantially very larger than the sensing capacitance; as a result the sensor's output tends towards values of decreasing dynamic range. Therefore after a cleaning procedure, it is required to cancel this offset and accordingly improve the output dynamic range.

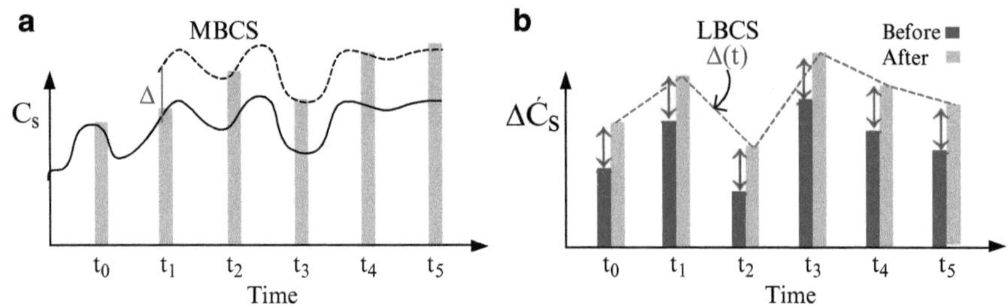

Fig. 4.1 Simplified illustration of the effect of offset error on (**a**) a MBCS and (**b**) a LBCS

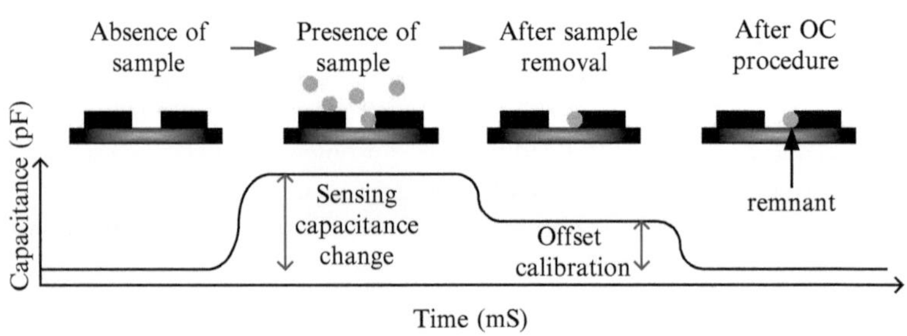

Fig. 4.2 An illustration of offset cancellation (OC) of a LBCS

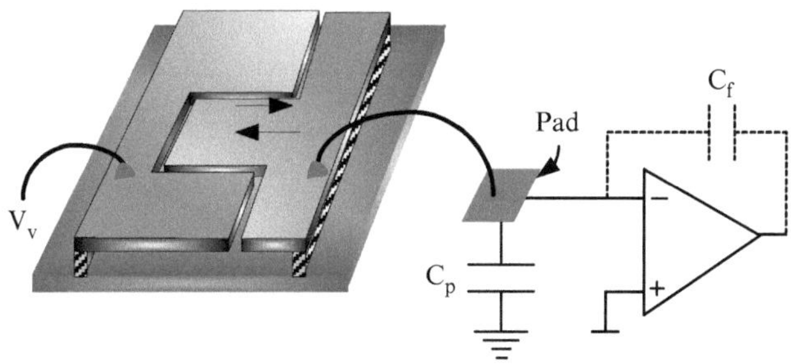

Fig. 4.3 Cancellation of parasitic capacitance in a MBCS

4.1.3 On-Chip Sensing Electrodes

The microfabricated electrodes for a MBCS are connected to CMOS circuitry through wire bonding or flip chip techniques (Fig. 4.3). The large parasitic capacitance (C_p) of the order of a few pico Farads (pF) created under the pad should be canceled with further circuitry (e.g. opamp with negative feedback). As was previously mentioned, the sensing electrodes of a LBCS can directly be realized above a CMOS chip and they are internally connected to an interface circuit without any extra bonding pad. LBCSs will often feature a large array of very small sensing electrodes which creates very small parasitic capacitances of the order of a few femto Farads (fF).

4.1.4 Measurement Time

A LBCS has sufficient time prior to and after introducing sample on the electrodes ($\Delta t1$, $\Delta t2$, Fig. 4.4a) while a MBCS for many applications including vibration monitoring should be able to measure capacitance variation over a very short period of time (δt) as shown in Fig. 4.4b. On other words, a LBCS can be implemented through a low speed circuitry based on the average of measured capacitances over a long duration of times $\Delta t1$, $\Delta t2$, but a relatively high speed and low noise capacitance measurement technique is required to make a MBCS.

4.1.5 RC Model Sample

As already mentioned, a MBCS is often encapsulated in a vacuum chamber and consequently a pure capacitance variation can be seen between electrodes, while biological or chemical samples above an electrode are modeled with a capacitance and resistance in series. Based on this important issue, the performance of dedicated conventional interface circuitry in a MBCS might be deteriorated when it is exposed

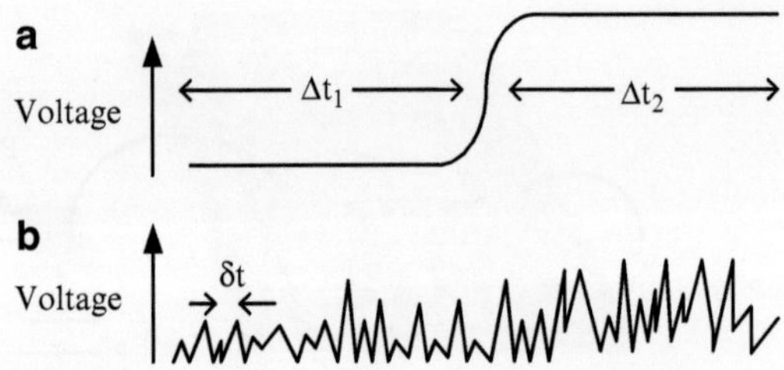

Fig. 4.4 Measurement times in (**a**) LBCS and (**b**) MBCS

to a biochemical samples. A LBCS allows the extraction of sensing capacitance from such a sample using a RC equivalent model.

In the next sections, the main capacitive techniques for LoC applications are introduced and the corresponding practical considerations will thereafter be discussed.

4.2 LBCS Methods

In this section, some very recently published methods involving LBCS are described. These methods include switched capacitor (SC), time constant, charge amplifier, charge based capacitance measurement (CBCM) methods.

4.2.1 SC-Based Interface Circuit

To date several SC interface circuitries have been reported for the measurement of acceleration, position, flow rate, pressure, humidity and strain [241–249]. Figure 4.5 shows a second order switch SC sigma delta (ΣΔ) modulator which was characterized as a fully differential capacitive interface sensor. By applying the reference voltages (V_{ref}) and clock pulses (Φ's) and by using an integrating capacitor (C_{int}) and other capacitors, the subtraction of sensing (C_S) and reference (C_R) capacitances is converted to digital output (y) as seen in Fig. 4.5. As this measurement technique relies on a differential readout voltage between sensing and reference capacitors, all the parasitic issues including temperature drift and aging affect the reference and sensing capacitors equally to the same way and therefore cancel out.

This circuitry can effectively measure the pure sensing capacitance for many applications such as chemical gas sensor [73]. As already described in Chapter 2, for gas sensor application, the polymer layer, on the top of electrodes assures that the effect of gas concentration causes a change in dielectric constant and subsequently in the capacitance of sensing electrodes.

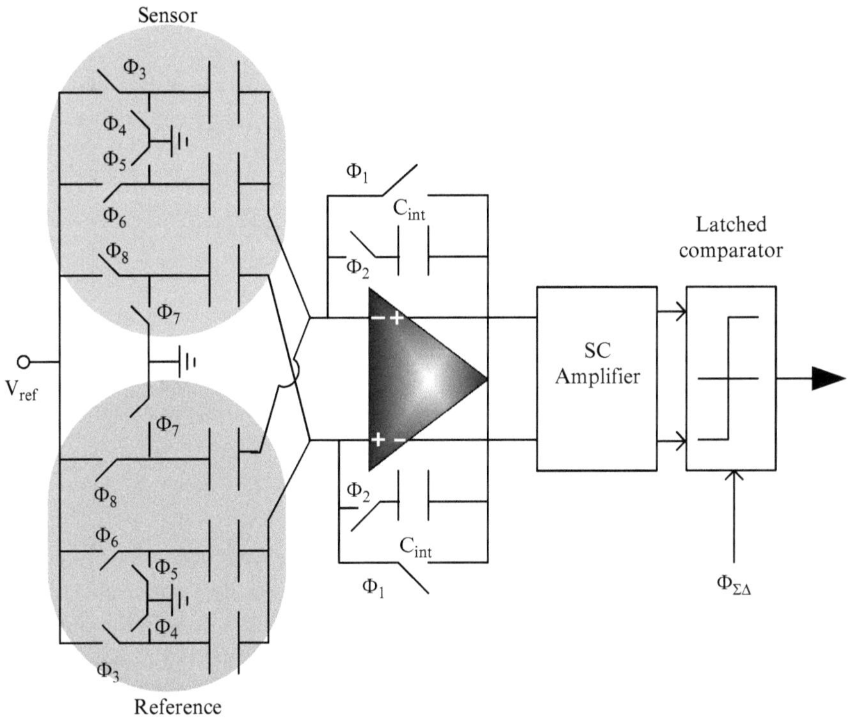

Fig. 4.5 Capacitive interface circuit through conventional switched capacitor topology

4.2.2 Time Constant Method

The measurement of capacitance in series with an unknown resistance (R) can be performed using the circuitry illustrated in Fig. 4.6a. As shown in this figure, the capacitance is charged and discharged when the applied pulse with amplitude of V_P and period of T becomes positive and negative respectively. If the time constant $\tau = RC < T/2$, the maximum value of output voltage V_C is then equal to $V_P(1 - e^{(-T/2RC)})$. By measuring $V_C(T/2)$ the capacitance value C can be extracted from this nonlinear relation (Fig. 4.6b). However, if $T/2 \gg \tau$, this measurement technique is not applicable, because $V_C(T/2) \approx V_P$ and is consequently independent of C (Fig. 4.6c). In order to avoid these problems and limitations, it is required to apply some modifications to the circuit topology.

A voltage comparator can be used to detect the voltage value of V(t) once it reaches $|V_{ref+}| = |V_{ref-}|$ which is smaller than $|V_{P+}| = |V_{P-}|$ As shown in Fig. 4.6d, a reference current (I_{ref}) can be applied to the electrodes in push or pull mode through switches SW1 and SW2. This therefore results in a transient voltage waveform on the sensing electrodes similar to V(t) in Fig. 4.6c. Once, this voltage becomes higher than a reference voltage (V_{ref}), the directions of reference

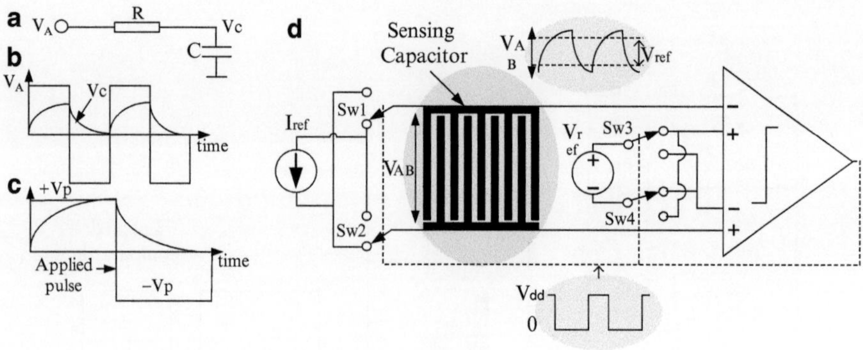

Fig. 4.6 Schematic of capacitive readout circuits based on time constant method, (**a**) RC equivalent circuit, (**b**) charging and discharging signals, (**c**) the charging/discharging signal of an (**d**) integrated circuit topology using time constant method

current and voltage are reverted simultaneously through SW1–SW4. The time constant of this sensing electrode is dominated by the capacitive electrode-solution interface. Consequently, the frequency of generated clock pulses in the output of comparator depends on C. This method was successfully applied by Stagni et al. for DNA detection using 0.18 μm CMOS process [72].

4.2.3 *Capacitive Inverter Amplifier*

The sensing capacitance Cs can be measured using a capacitive inverter amplifier as shown in Fig. 4.7a. By applying a voltage pulse V_p (Fig. 4.7b), the output voltage (V_{out}) of the opamp can be obtained from Eq. 4.1

$$V_{out} = \frac{V_1 \bullet C_S + V_2 \bullet C_R}{C_{in} \| C_P} \tag{4.1}$$

where C_R is a reference capacitance and node "A" is virtually grounded. As the parasitic capacitance C_p standing in between the nodes "A" and "O" is in parallel with C_{in}, the demonitor can be expressed simply as (C_{in} + C_p) consequently this results in a lower amplification factor. The dynamic range of this circuit depends on C_S, which is equal to the summation of a constant value C_{S0} and the variable sensing capacitance ΔC_S. The dynamic range of the output voltage of this design is increased by an appropriate value of C_R ($\approx C_{S0}$) [74].

The parasitic capacitance C_{P1} is virtually grounded and the integrating capacitance C_{in} resets once each period of applied clock pulses V_1 and V_2. In fact, the subtraction of C_S and C_C appears in the numerator of Eq. 4.1. This sensing method which was successfully employed for bioparticle localization [74], can be realized using a low complexity circuitry as shown in Fig. 4.8.

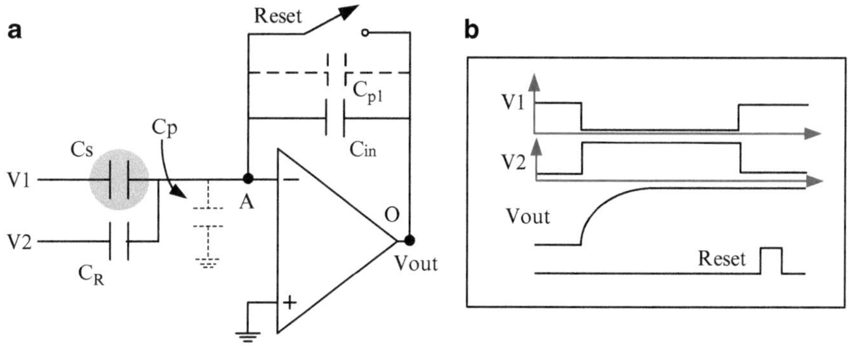

Fig. 4.7 Schematic of a conventional inverter amplifier for capacitive detection: (**a**) circuit topology and (**b**) the applied clock pulses and corresponding output voltage

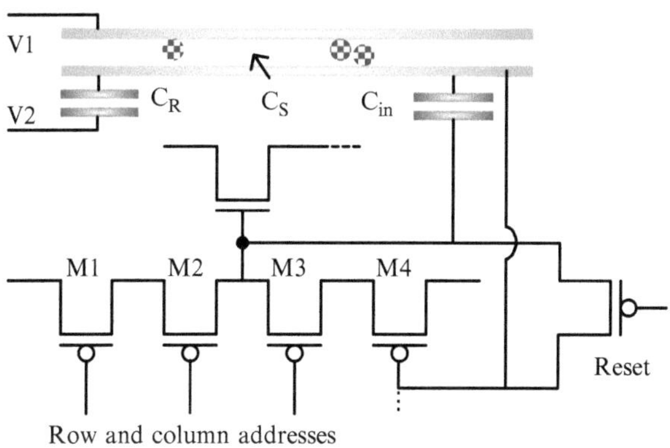

Fig. 4.8 Schematic of interface circuit using inverter amplifier technique

In this figure, the p-channel transistors M1–M4 are integrated to create an opamp which is selected by applying appropriate row and column addresses. Another switch is used to reset the feedback capacitance (C_{in}). Similar the schematic shown in Fig. 4.7, V_1 and V_2 are applied on C_S and $C_{R.}$

Actually, this capacitive amplifier circuitry can be considered a follow up to the work of previously reported capacitive sensor method for fingerprint recognition purposes [250, 251]. As shown in Fig. 4.9, the sensing capacitance C_S is embedded in the feedback loop instead of C_{in} in Fig. 4.7.

By applying a current pulse I_{input} instead of voltage V1 as in Fig. 4.7, the capacitance variation on the sensing electrodes attached by the ridge of the finger's skin is sensed. It is worth to mentioning that the main problem associated with capacitive sensors for fingerprint applications is the difficulty of sensing very small capacitances that are in the order of a few femto Farads.

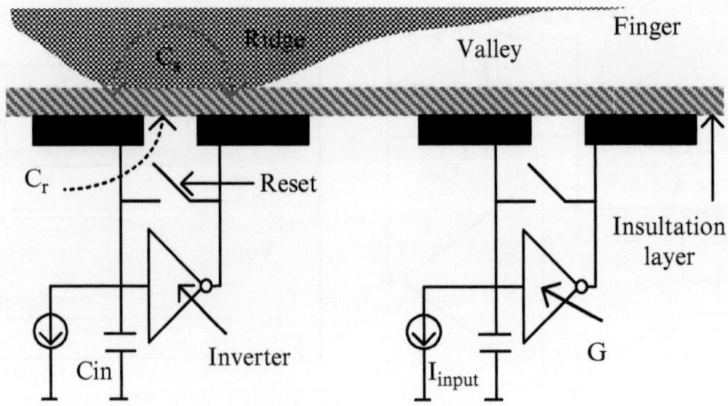

Fig. 4.9 Illustration of a capacitive fingerprint recognition [250]

The output voltage variation is inversely proportional to the capacitance variation as represented in Eq. 4.2,

$$\Delta V_0 \approx \frac{\delta Q}{C_r} \tag{4.2}$$

where δQ is the charge taken by C_r from input current. It should be mentioned that it is assumed that the gain amplifier $G \gg 1$ and $C_p \ll Cr$ [251] where C_p is a parasitic capacitance which is due to the electric fringing field and is parallel with C_r. Based on this relation, the small variation of C_r is sensed, thus increasing the noise immunity and the robustness of the readout structure. In another effort, a floating capacitor technique was used to detect the ridge of fingerprint as shown in Fig. 4.10. As can be seen in this figure, three transistors M1–M3 charge and discharge the nodal parasitic capacitances C_{N1} and C_{N2}. While switches M1 and M3 are turned on and M2 is turned off, N1 and N2 are connected to V_{dd} and V_{SS} respectively. In another phase of sensor operation, M2 is turned on, while M1 and M3 are off. In this phase, the charges between C_{N1} and C_{N2} are redistributed and the output voltage is obtained from the following equation.

$$V_{out} = \frac{(C_{N1}+C_S)V_{dd}+C_{N2}V_{SS}}{C_{N1}+C_{N2}+C_S} \tag{4.3}$$

By minimizing the nodal parasitic capacitances, the variation of output voltage can be increased significantly. For this purpose, the fringe capacitances between the sensing electrode and the substrate can be cancelled by shielding the sensor circuitry using Metal 2 with a potential that tracks the sensing electrode potential using a unity-gain buffer (Fig. 4.10).

This circuit topology was thereafter employed for cell monitoring where the passivated electrodes are exposed to living cells for monitoring the living/mortality of cells and for tracking cancer cell proliferation as described in Chapter 2 [142].

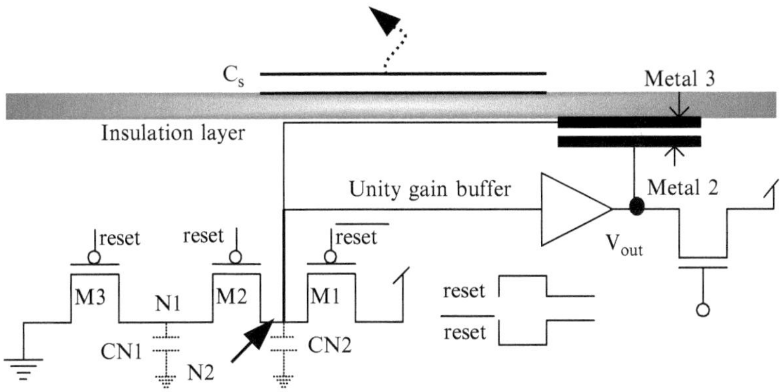

Fig. 4.10 Schematic of capacitive interface circuit measuring the floating capacitor C_S

4.2.4 *CBCM Methods*

Charge based capacitance measurement (CBCM) method was invented as an accurate technique for the characterization of interconnects capacitance in deep submicron CMOS ICs [252, 253]. Figure 4.11 shows the principle of CBCM in which two signal pulses (Φ1 and Φ2) are applied to two pairs of n/pMOS transistors in order to frequently charge and discharge the interconnect capacitance C_L. Based on this method, the average of charging (or discharging) current is proportional to C_L as represented in Eqs. 4.4 and 4.5.

$$I - I' = I_{net} \tag{4.4}$$

$$I_{net} = f \bullet V_{dd} \bullet C_L \tag{4.5}$$

where I, I′ and I_{net} are the average of charging current on both arms of the CBCM circuit and the subtractions of these currents respectively (Fig. 4.11). Also, V_{dd}, f, and I_S are the power supply voltage, frequency and DC charging currents respectively. When similar devices are used on both sides of the CBCM structure, the effect of parasitic capacitances associated with M1–M2 is removed from I_{net}, and consequently based on the linear relation shown in Eq. 4.5, CBCM allows a sub femto Farad (fF) capacitance measurement [252].

Several modifications have been applied to CBCM structure in order to increase the accuracy of CBCM [254–256] or to employ CBCM in other applications including the characterization of nonlinear capacitors [257, 258] and measurement of bias-dependent capacitance [259] Even though CBCMs were originally proposed for the measurement of static capacitance in deep CMOS chip, this measurement technique has recently attracted attentions for the measurement of dynamic capacitance atop CMOS chip. Based on this, CBCM offers low complexity and highly precision methods to detect the capacitance variation of sensing electrode realized

Fig. 4.11 Illustration of CBCM method

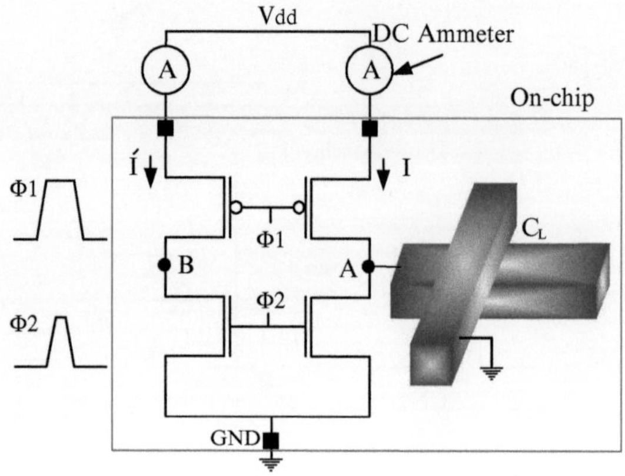

on the top most metal layer of CMOS chip. The new advantage of CBCM was successfully exploited for biological and chemical applications such as DNA detection, cellular activity, bacteria growth and organic solvent monitoring [164, 260–263]. Furthermore modifications should be applied to UC Berkeley CBCM structure in order to make a generic capacitive sensing system suitable for such applications. Most importantly, it is required to design a compact, low-noise and low power consumption circuitry instead of using external current measurement device in order to determine the average charging current of sensing capacitance. For sensing applications, CBCM is considered as an interface circuit which converts the capacitance values to charging current. In the next section, we describe various core-CBCM circuit topologies which are suitable for LoC applications.

4.3 Core–CBCM Interface Circuit

In this section, various core-CBCM circuitries with different levels of complexity are described for integrated capacitive sensor LoCs.

4.3.1 Principle of CBCM for Sensing Applications

As already mentioned, the total capacitance of sensing electrode C_S is equal to the summation of the variable sensing capacitance ΔC and constant capacitance C_0. As shown in Fig. 4.12a, the CBCM structure is a good candidate to precisely remove C_0 from C_S

As shown in this figure two similar capacitive electrodes C_R and C_S are connected to nodes A and B. C_R and C_S play the role of reference and sensing electrodes respectively. Based on CBCM method, the subtraction of charging currents I_R and I_S is proportional to the subtractions of the capacitances standing

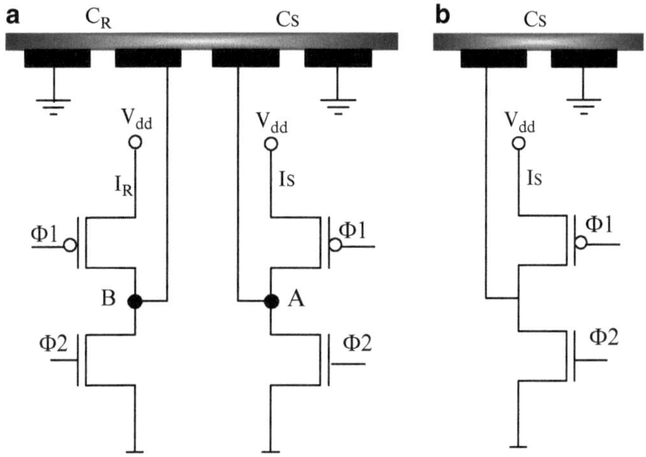

Fig. 4.12 CBCM capacitive sensor: (**a**) four transistors and (**b**) two transistors

on nodes A and B. Therefore, by assuming that similar sensing electrodes and transistors are present on both sides of the CBCM structure, the sensing capacitance can be obtained through the following equations.

$$I_R = f \bullet V_{dd} \bullet C_R \tag{4.6}$$

$$I_S = f \bullet V_{dd} \bullet C_S \tag{4.7}$$

By subtracting the Eqs. 4.1 and 4.2, the following relation is obtained.

$$I_S - I_R = \mathrm{f} \bullet V_{dd} \bullet (C_S - C_R) \tag{4.8}$$

Therefore, the differential current ΔI is proportional to sensing capacitance ΔC as shown in Eq. 4.9.

$$\Delta I = f \bullet V_{dd} \bullet \Delta C \tag{4.9}$$

4.3.2 Two Transistors CBCM Sensor

The capacitance changes of a single electrode can be measured in two steps. As shown in Fig. 4.12b, one electrode is grounded while another is connected to the drain sides of a n/pMOS pair. In the first step, the charging current I_1 is measured using a DC Ammeter (A) while the channel is empty. Therefore, the combination of all parasitic capacitances related to MOS transistors and electrodes is obtained from Eq. 4.10

$$I_1 = f \bullet V_{dd} \bullet C_0 \tag{4.10}$$

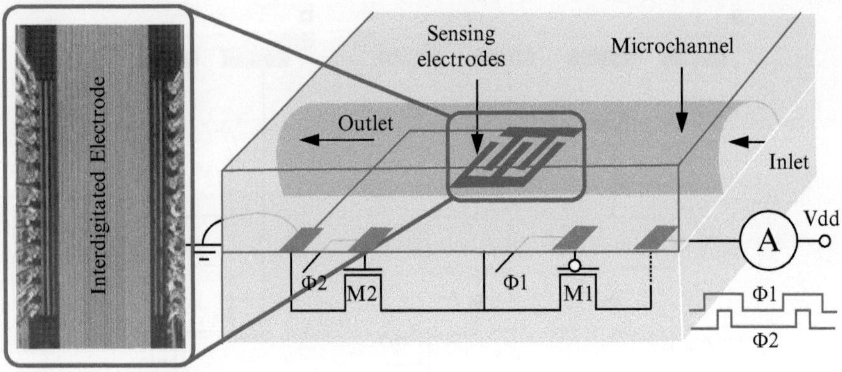

Fig. 4.13 Hybrid microfluidic/CMOS capacitive sensor using a two-transistor CBCM (A – Ammeter)

where *f*, C_0 and V_{dd} are the frequency of pulse signals, the parasitic capacitance and the power-supply voltage respectively. At the second step, the solution is injected into the microchannel and the charging current (I_2) is measured. As given in Eq. 4.11, this current is proportional to the combination of C_0 and the capacitance variation ΔC.

$$I_2 = f \cdot V_{dd} \cdot (C_0 + \Delta C). \tag{4.11}$$

Therefore, C_0 and ΔC could be obtained using Eqs. 4.10 and 4.11 for a given value of *f*. A 0.18 CMOS chip was fabricated using the above mentioned two-transistors CBCM circuit for organic solvent monitoring [66, 264]. Figure 4.13 shows the schematic of a hybrid Microfluidic/CMOS capacitive sensor using such two-transistor CBCM method. This hybrid system consists of a CMOS chip with a large interdigitated electrode connected to the drains of transistors. Off-chip clock pulses are applied to this chip and a DC ammeter is used to measure the charging current.

The differential charging current ($I_1 - I_2$) versus frequency for four solvents are obtained and given in Fig. 4.14 [66]. These solvents are dichloromethane, acetone, methanol and deionized water (DW). The dielectric constants of these solutions are 9.1, 20.7, 30.4 and 80.4 respectively at 69°F. It should be mentioned that the conductivity of dichloromethane as non-polar solvent is almost zero and the low conductivities of DW (0.04 μS/cm) and acetone (0.02 μS/cm) can be discounted. In this work, V_{dd} remained at its maximum value (1.8 V) and the frequency was varied up to four decades in order to demonstrate the linearity and viability of Eq. 4.11 for dynamic capacitance characterization. For this wide frequency range, the measurement results are revealed in logarithmic scales. The parallel straight lines in Fig. 4.14a are in agreement with Eq. 4.12, which results from the combination of Eqs. 4.10 and 4.11.

$$\log(I_2 - I_1) = \log f + \log(V_{dd} \cdot \Delta C) \tag{4.12}$$

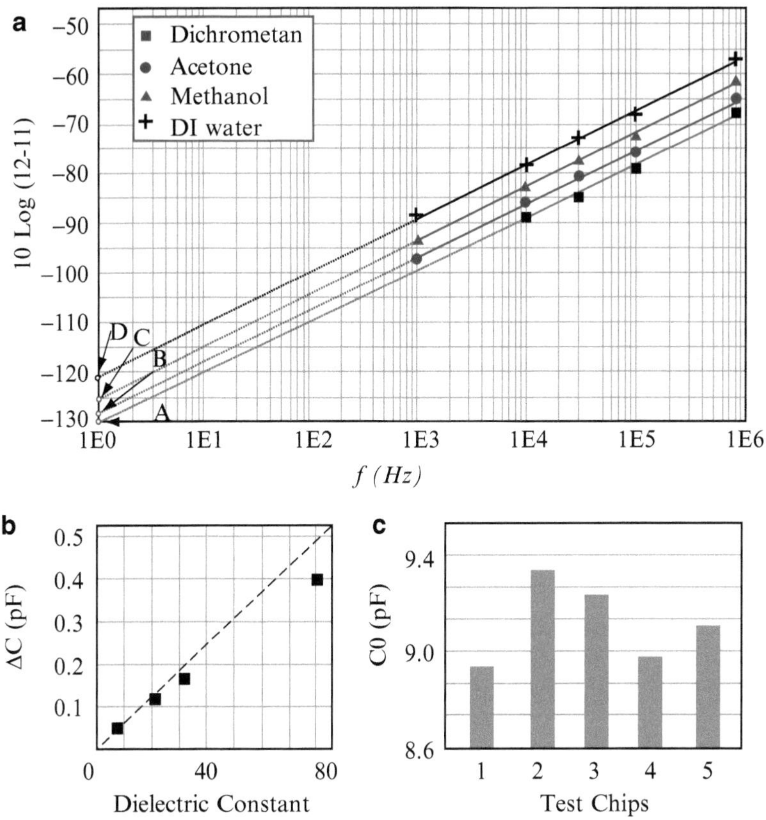

Fig. 4.14 Measurement results: (**a**) differential current versus frequency for four different solutions, (**b**) the capacitance variation of chip versus dielectric constant, and (**c**) the parasitic capacitance of five different chips [66]

Therefore, the capacitance variation (ΔC) associated with each solution can be extrapolated from the x-intercept of these curves (Fig. 4.14b). In fact, in this logarithmic demonstration, for different solutions the width of origin (A, B, C and D) is different while the slopes of these curves are the same. Similar measurements were performed to extract C_0 for five chip samples when no solution has been introduced on the electrodes (Fig. 4.14c). The difference between the extracted C_0 from different chips could be due to non-uniformity of pad-etching and/or CMOS process tolerance.

Finite element based CAD tools such as FEMLab are conventionally applied for assessment of such capacitance changes. However, in addition to physical dimensions, the modeling of such applications requires a lot of preliminary data, such as physicochemical and processing data which are seldom available. Thus, an experimental characterization method that could carry out analyses accurately without relying on previously available data is highly desirable. It is believed that

CBCMs are the best alternative for the characterization of sensing electrodes. It is obvious that two-transistor CBCM circuit can be used as the frond-end stage of a capacitive sensor and further circuitries should be incorporated to implement a fully integrated capacitive sensor. We will discuss the relevant technique in the following sections.

It is worth pointing out that this chip can also be used in a simple method for the characterization of capacitive LoC sensors. Through this characterization procedure, the input capacitance variation and contributed parasitic capacitances are extracted.

This extracted data could be used as a simplified model for design and optimization of on-chip circuits dedicated to LoC applications.

4.3.3 Opamp-Based Integrator Incorporated with CBCM Sensor

The first step toward a core-CBCM integrated sensor is the measurement of the charging current. A good alternative is an opamp-based integrator similar to the one shown Fig. 4.15a [192]. As shown in this figure, the discharging current can be conducted through node A. By applying clock pulses, the average current in the discharging half-period results in an output voltage. The value of voltage can be calculated from the following relation [192].

$$V_{out} = (-V_{dd} + \frac{T}{C_{in}} \bullet I_{dc} / 2) - \frac{C_S}{C_{in}} \bullet V_P \tag{4.13}$$

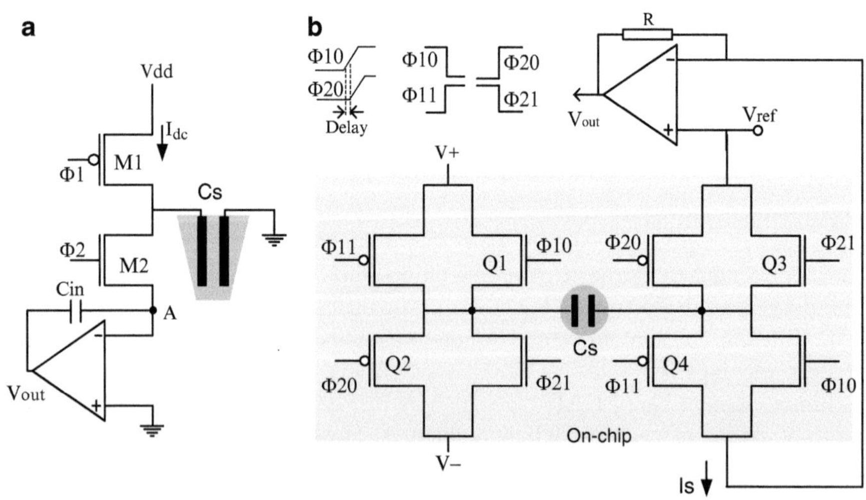

Fig. 4.15 A core-CBCM capacitive DNA detection: (**a**) Off chip realization and (**b**) on chip realization of new CBCM structure

where C_{in}, V_P and $T = 1/f$ are the integrating capacitor, the amplitude and the period of the applied clock pulses respectively. I_{dc} is also the quiescent current. Thanks to negative feedback established by feedback capacitance C_{in} the opamp keeps a constant voltage on the source of nMOS transistor (M2). This method has successfully been applied for DNA detection using discreet devices by Guiducci et al. [192]. As a follow up to this work, they successfully realized the same methodology of DNA sensing on a CMOS chip [193]. In that work, a CBCM structure was implemented on a chip and the gold electrodes were realized through a sputtering procedure. Figure 4.15b shows a modified CBCM structure with a sensing capacitor in between the drains of the transistors and four clock pulses Φ10, Φ11, Φ20 and Φ21 along with additional p- and n-channel transistors (Q1–Q4) in order to exploit the full voltage range. In this bridge topology, Φ20 is applied with a short delay with respect to Φ10. Additionally, an off-chip circuit including an opamp and a resistor (R) was employed to convert the current signal (I_S) into a readable voltage by a PC [72], with appropriate values of DC biases V^+ and V^-.

As already mentioned, in order to improve the dynamic range of the system and consequently increase the sensitivity, it is required to subtract the constant value of sensing capacitance. In this direction, let us describe some specific integrated core-CBCM circuitries.

4.3.4 *Differential Current CBCM Techniques*

The discharging currents can simply be subtracted in a circuit topology like the one shown in Fig. 4.16a. This differential structure CBCM operates in two phases by non-overlapping clocks -Φ1, –Φ2, Φ1, and Φ2. In the first phase, M2 and M3 are

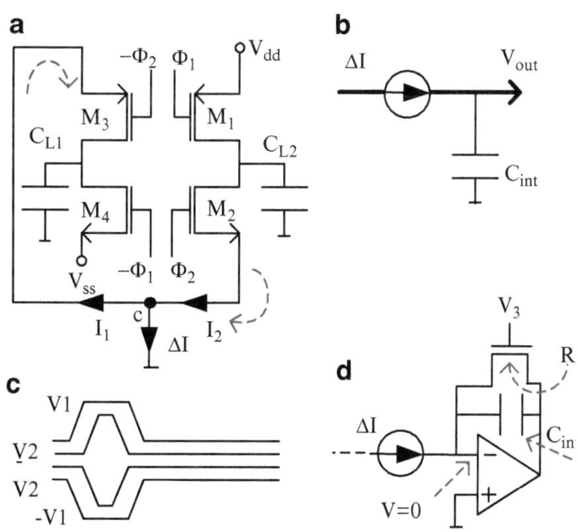

Fig. 4.16 Differential CBCM structure: (**a**) current differtial circuitry with corresponding waveforms, (**b**) simple integrator, and (**c**) opamp based integrator

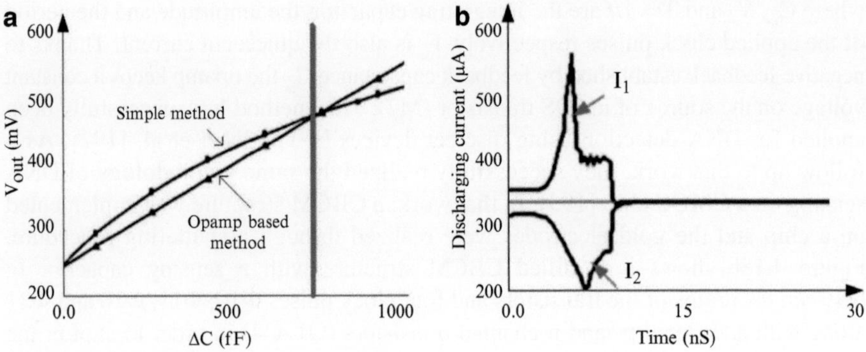

Fig. 4.17 Simulation results of the circuitries shown in Fig. 4.16: (**a**) comparison of linearity between the simple capacitive integrator and opamp based integrator and (**b**) mismatch current error

turned on while M1 and M4 are turned off. The discharging currents of C_{11} and C_{12} are subtracted in node C. In the second phase, ΔI = 0 because the M2 and M3 are closed. ΔI should be integrated through a single capacitor or opamp-based integrator as shown in Fig. 4.16b, c respectively.

As shown in Fig. 4.17a. the opamp-based integrator allows for a linear relation to exist between the output voltage and differential capacitance as compared to the simple capacitive integrator method. In this simulation, a conventional two-stage opamp [265] was used. The key parameters of this opamp are that its unity gain bandwidth (UGB), DC voltage gain and slew rate with values 25 MHz, 83 dB and 14 V/μS, respectively [39].

The main problem of this circuit topology is error associated with ΔI. The simulation results (Cadence, Spectres) shows that I1 and I2 in Fig. 4.16a are not the same even if similar devices are used in this topology. This error is due to mismatch between PMOS and NMOS transistors.

4.3.5 Current Mirror Integrated with CBCM Structure

A simple current mirror (M_3–M_4) can be integrated with a CBCM structure in order to measure the charging current instead of using an off-chip ammeter, as shown in Fig. 4.18a. This current mirror can sense and amplify the CBCM charging current (I_S) with a current gain equal to A_I. As shown in Fig. 4.18b, once Φ1 is low and the voltage on C_S starts rising rapidly, the following relation is established in between the charging current through Cs, drain current of M_3 and charging current through C_{in} to this capacitance to voltage converter (CVC) unit.

$$A_I \cdot C_S \cdot \frac{dV_S}{dt} = A_I \cdot K_X \cdot (V_{gs} - V_{TP})^2 = I_S \tag{4.14}$$

In this relation, K_x depends on the parameters of the process. Also, V_{gs} and V_{TP} are the gate source voltage and the threshold voltage of the p channel MOSFET respectively. Also, the switch-on and off resistances of M_1 and M_2 have been discounted.

By substituting $V_{dd} - V_S$ instead of V_{gs} in this equation and also by assuming $V_S = 0$ at t = 0 (discharging) when Φ1 and Φ2 are high, V_S can be expressed by

$$V_S = (V_{dd} - V_{TP}) - \frac{(V_{dd} - V_{TP})C_S}{K_x(V_{dd} - V_{TP})t + C_S}. \tag{4.15}$$

Therefore, $I_S = I(C_S,t)$ can easily be obtained based on the combination of the above mentioned relations

$$I(C_S,t) \approx \frac{[(V_{dd}-V_{TP})C_S]^2 K_x A_{I1}}{[K_x(V_{dd}-V_{TP})t+C_S]^2}. \tag{4.16}$$

where A_{I1} is the gain of current mirror (M3–M4). This time-variant current signal can be considered as the equivalent model of circuits shown in Fig. 4.18b with a capacitive load C_{in}. By combining Eqs. 4.15 and 4.16 and the relation of integrating capacitor,

$$V_{out} = C_S \bullet \frac{d(I_S)}{dt} \tag{4.17}$$

V_{out} can be ideally given by Eq. 4.18 for $t \gg 0$

$$V_{out} = \frac{C_S}{C_{in}} \bullet \frac{W_4}{W_3} \bullet \frac{L_3}{L_4} (V_{dd} - V_{TP}). \tag{4.18}$$

where W_3/L_3 and W_4/L_4 are aspect ratios of M_3 and M_4 respectively. As $C_S = \Delta C + C_0$, it is obvious, the dynamic range of V_{out} remains limited due to$\Delta C \gg C_0$. In order to extract the sensing capacitance ΔC from C, two different solutions have already been reported [266, 267]. These solutions are discussed in the following sections.

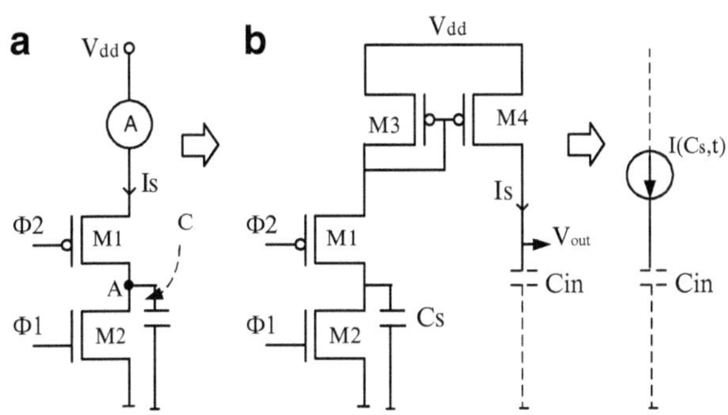

Fig. 4.18 Core-CBCM sensor: (**a**) CBCM structure and (**b**) on-chip CVC unit

4.3.5.1 Differential Voltage Technique

The first solution is to generate a reference voltage V_R by employing a replica of the circuit shown in Fig. 4.18b with C_R instead of C_S. A differential voltage amplifier can be used to subtract the V_S from V_R as shown in Fig. 4.19. Therefore, by assuming a symmetrical circuitry with similar transistors ($M1 \equiv M1_R$, $M2 \equiv M2_R$, $M3 \equiv M3_R$, $M4 \equiv M4_R$) and the same capacitance value for the integrating capacitor ($C_{in} \equiv C_{in\text{-}R}$) the output voltage can be derived from the Eq. 4.19

$$V_{out} = \frac{\Delta C}{C_{in}} \bullet \frac{W_4}{W_3} \bullet \frac{L_3}{L_4} \bullet (V_{dd} - V_{TP}) \bullet A_V \tag{4.19}$$

where A_V is the voltage gain of the differential amplifier. This technique was previously reported for on-chip particle detection [268, 269].

In Fig. 4.19, SW1 and SW2 are used to reset the integrating capacitors C_{in} and C_{in_R} before beginning the charging phase. During this phase, a problem is arisen from the voltage differential amplifier. In fact, the higher sensitivity and consequently higher output voltage of each CVC unit, pushes the voltage differential amplifier into the nonlinear region, thereby derogating the resolution of the capacitive sensor.

A simple solution to avoid both the above mentioned problems is to subtract the charging currents resulting from the sensing and reference capacitances prior to injection in integrating capacitors. We will progress forward on this topic by describing a differential current method and further digital readout circuitry as seen in the next section.

4.3.5.2 Differential Current Technique

As already mentioned, the sensing capacitance ΔC value is lower than C_S therefore, a low dynamic range output voltage is expected for each CVC unit. A solution for this

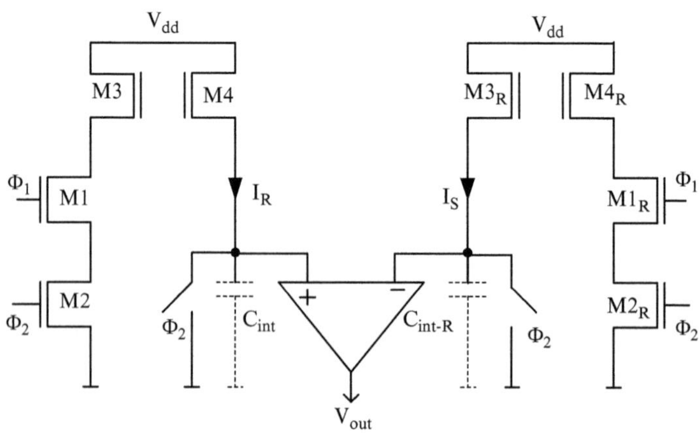

Fig. 4.19 A schematic of a circuit used in the differential voltage core-CBCM technique

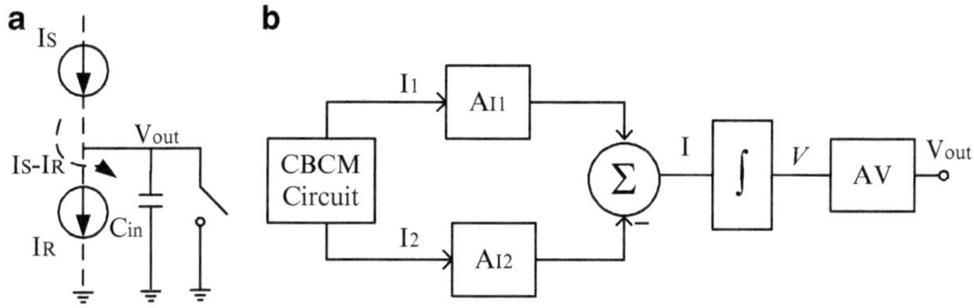

Fig. 4.20 Differential current method: (**a**) simplified circuitry and (**b**) system level

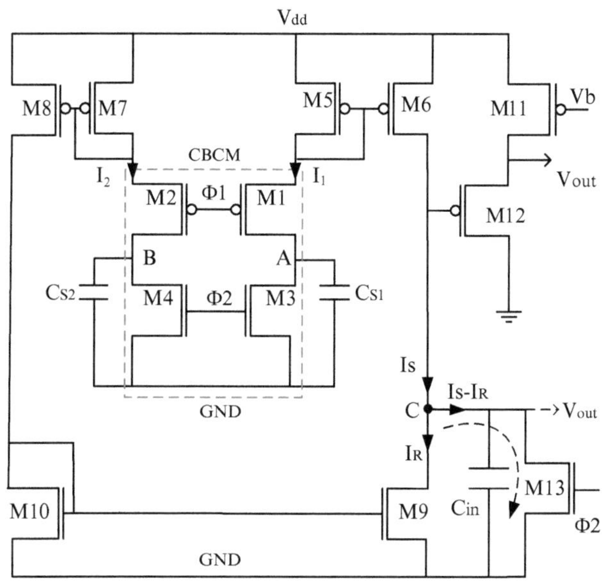

Fig. 4.21 Proposed capacitive interface circuit topology

problem is to generate a reference current I_R by employing a replica of CVC unit with C_R instead of C_S. The differential current $I_S - I_R$ is injected into C_{in} (Fig. 4.20). Thanks to its symmetry and differential operation, V_{out} is expressed by Eq. 4.20 [64]

$$V_{out} = \frac{\Delta C}{C_{in}} \bullet \frac{W_4}{W_3} \bullet \frac{L_3}{L_4} (V_{dd} - V_{TP}) \tag{4.20}$$

The differential current based circuit and its equivalent diagram are depicted in Fig. 4.20a, b respectively. As shown in these figures in the differential current based circuit topology shown in Fig. 4.21, two current mirrors (M5–M6 and M7–M8) amplify (A_{I1} and A_{I2}) the charging currents I_1 and I_2 and the third current mirror (M9–M10) is employed to transfer I_2 to node C. The difference between the DC components of I_1 and I_2 is injected into C_{in}. The output voltage of the amplifier is also buffered in the output stage of circuit through M11–M12 transistors. M13 is used to reset C_{in}.

Ionic Conductive Solutions

So far in this chapter, we assumed that a pure dielectric or non-conductive sample is introduced to the sensing electrodes. Now let us apply a conductive solution to the electrodes and then obtain the output voltage for t > 0. By assuming a conductive solution, the electrical model of the sample becomes a parallel resistor R and a capacitor C. Also, as the sensing electrodes are covered by a passivation layer atop the CMOS chip, another capacitance C1 is created across this layer. Therefore the output voltage can be obtained from Eq. 4.21 [221].

$$V_{out} = \frac{C_1}{C_{in}} \cdot A_{I1} \cdot V_{dd} \cdot \left(1 - \frac{C}{C + C_1} e^{-\frac{t}{R(C_1 + C)}}\right) \cdot V_{off} \tag{4.21}$$

As seen in this equation, for $t \gg 0$, $V_{out} = A_{I1} \cdot C_1 \cdot V_{dd}/C_{in} + V_{off}$. In other words, by applying low frequency clock pulses, R does not affect V_{out} and V_{out} only is a function of C.

Differential Capacitive Sensor

If the reference and sensing capacitors are exposed to two different analytes, a comparison between the solutions can be performed using the core-CBCM sensors as shown in Fig. 4.22a. This important advantage of core-CBCM was reported for bacteria growth monitoring [210, 211]. For this application, as mentioned in Chapter 2, the reference electrode is exposed to LB medium while the sensing electrode is exposed to a bacterial solution. The electrical models of the analytes

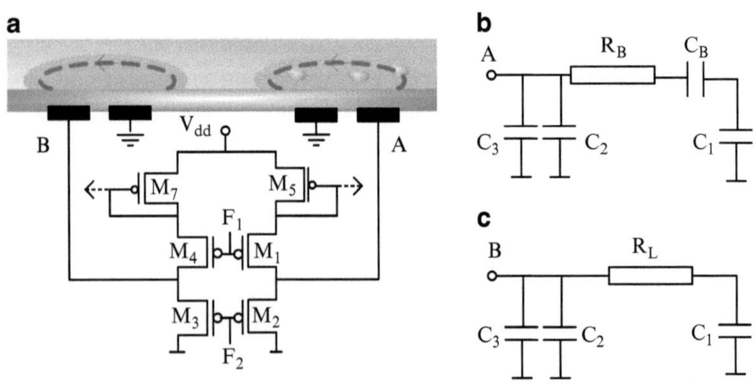

Fig. 4.22 Differential capacitive sensor: (**a**) interface circuit and equivalent model of sensing electrodes exposed to (**b**) a bacterial solution, and (**c**) an LB medium

above the reference and sensing electrodes are also shown in Fig. 4.22b, c. Based on these models, R_L and R_B represent the resistive properties of L_B and the bacterial solution. By applying the clock pulses with lower frequencies $f = 1/T$, V_C become less dependent on R_B and R_L as expected from the following equations [31].

$$V_C = \frac{1}{T}\int_0^T \frac{V_{dd} - V_{TP}}{D_{in}} \bullet A_I \bullet \left(\frac{e^{\frac{-2t}{R_B C_I}}}{R_B} - \frac{e^{\frac{-t}{R_L C_B \| (C_I/2)}}}{R_L} \right) dt \Bigg|_{T>>0} + V_{off} \tag{4.22}$$

$$= (V_{dd} - V_{TP}) \bullet A_I \bullet \frac{(C_I / 2 - C_B \| C_I / 2)}{C_{in}} + V_{off}$$

By assuming constant values for all parameters in Eq. 4.22 (except C_B), the output voltage will be a function of C_B only. Based on this discussion, the differential current CBCM sensor can effectively measure, the differential capacitive properties of two different solutions.

Sensor Characteristics

Linearity

The transient output voltage of the interface circuit was simulated by Spectre S in Cadence for different values of input sensing capacitances (C_{S1}). Based on these results shown in Fig. 4.23a, b, the output voltage was seen to be a linear function of input differential capacitance. Figure 4.23b also shows the sensitivity which was obtained using the following relation.

$$S = \frac{dV_{out}}{d(\Delta C_I)} \tag{4.23}$$

Approximately a uniform sensitivity is expected in the middle region of this figure

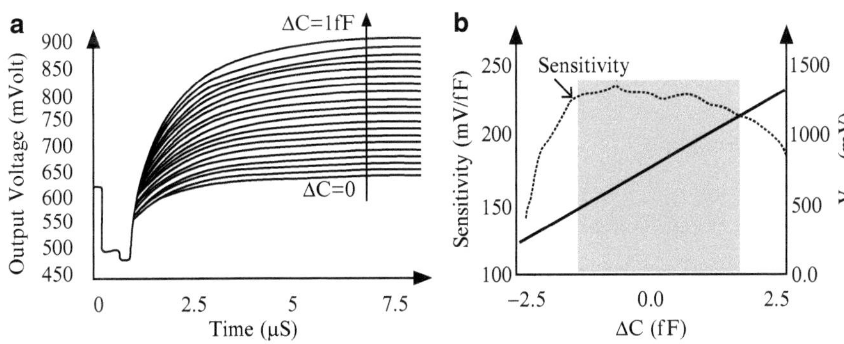

Fig. 4.23 Simulation results of circuit shown in Fig. 4.21: (**a**) output voltage versus time for different values of C_{S1} and (**b**) output voltage versus differential sensing capacitance ($\Delta C_S = C_{S1} - C_{S2}$)

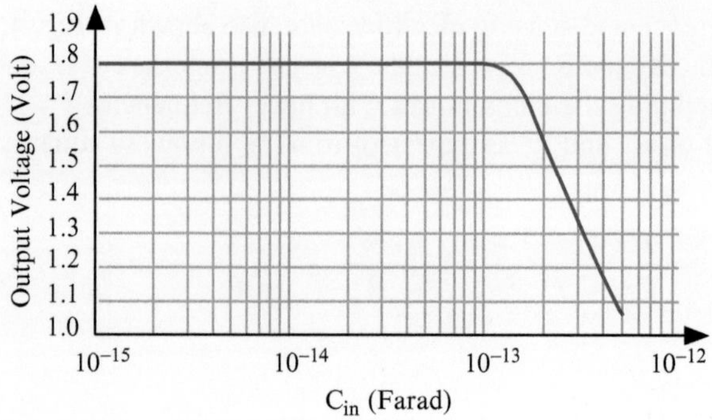

Fig. 4.24 Output voltage versus C_{in}

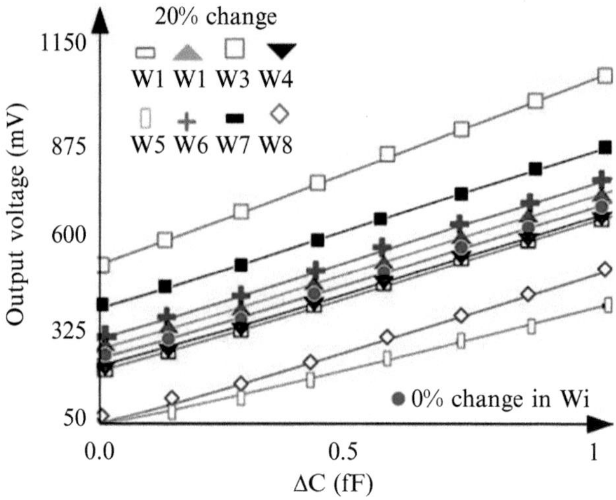

Fig. 4.25 Variation of simulated V_{out} versus ΔC for different mismatch errors (20% change of a transistor's width)

Integrating Capacitor

As illustrated in Fig. 4.24, the small values of C_{in} (less than 100 fF) do not increase the sensitivity as was expected in Eq. 4.23. This is due to the comparable parasitic capacitors on node "C".

Geometrical Issues of Transistors

As the stray capacitances of the transistor depend on its dimensions, a mismatch between the transistors in both sides of symmetrical topology shown in Fig. 4.25 may cause an offset in the output voltage. By assuming the same length (L = 1 μm) for all transistors, this effect has been studied using the simulations with SpectreS. As shown in this figure, 20% change of W_1, W_2 or W_5–W_8 significantly affects V_{off}

with no change in sensitivity. Of course, a calibration procedure can be applied to cancel this error by adjusting the current gain (M_3–M_4). These simulations have been carried out for a current gain of 10 with C_{in} = 1.2 pF.

Thermal Issues

The thermal voltage drift on V_{out} is another issue which is studied using simulations. The variation of V_{out} (t = 95 ms) versus temperature shown in Fig. 4.26. This figure indicates only a 33 mV deviation in V_{out} for a wide range of temperatures (0–120°C) where deviation is much less than ΔV_{out}. In reality however, the temperature variations of the fluid inside the channel atop the chip cannot surpass this voltage deviation.

However the differential current method allows for a linear relation to be drawn between the output voltage and input capacitance. Error sources resulting from geometrical mismatch, thermal drift or the remnant in the microchannel above the sensing electrode can cause an offset voltage in V_{out}.

As the measurement through the capacitive sensor for biological applications is performed in two different steps prior to and post analyte injection in the channel, the offset voltage which is independent of the sensing parameters (for instance the concentration of analyte which is under test), can be cancelled by subtracting the output voltage measured in two steps. Therefore, the offset voltage does not affect the linearity or the sensitivity of the system, but it may decrease the dynamic range of the system by pushing the output stage of the system into nonlinear and saturated states. For this reason, it is necessary to cancel the offset voltage as discussed in the next section.

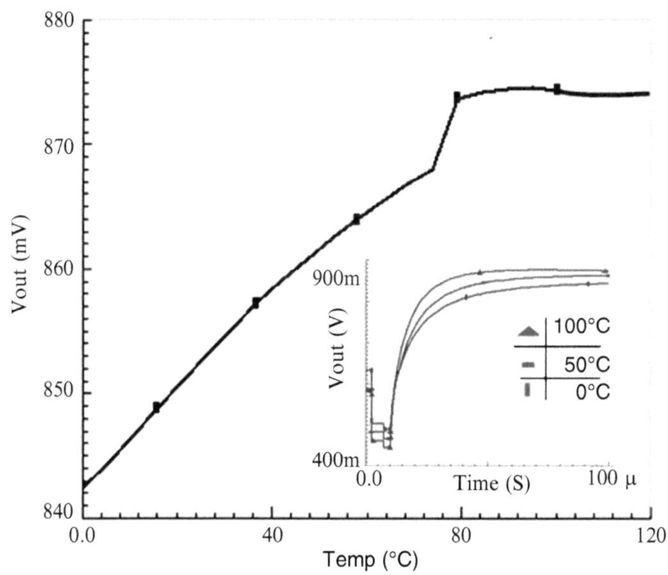

Fig. 4.26 Simulation results of V_{out} variations versus temperature. The transient responses for 0°C, 50°C and 100°C are shown in the inset

Calibration Technique

Let us assume V_{off} is the residual offset voltage mainly resulting from the transistor mismatches in the current mirrors, the CBCM structures, as well as the mechanical artefacts, the remnants in the microchannel, and the sensing electrode mismatches. The output voltage can be represented as in Eq. 4.24.

$$V_{out} = \frac{\Delta C}{C_{in}} \bullet \frac{W_4}{W_3} \bullet \frac{L_3}{L_4} (V_{dd} - V_{TP}) + V_{off} \quad (4.24)$$

Off-Chip Resistive Technique

The offset voltages can be cancelled by incorporating two off-chip potentiometers R_{P1} and R_{P2} as shown in Fig. 4.27a. These potentiometers are used to vary A_{I2} and consequently change the output voltage in order to cancel the effect of V_{off}. Based on the simulation results shown in Fig. 4.27b, the output voltage can be varied for different values of R_{P1} or R_{P2}, however, only a coarse adjustment of R_{P1} (or R_{P2}) can be done due to the nonlinear relation between V_{out} and these potentiometers. For this, a high resolution calibration procedure can be performed using digital circuitry as described in the next section.

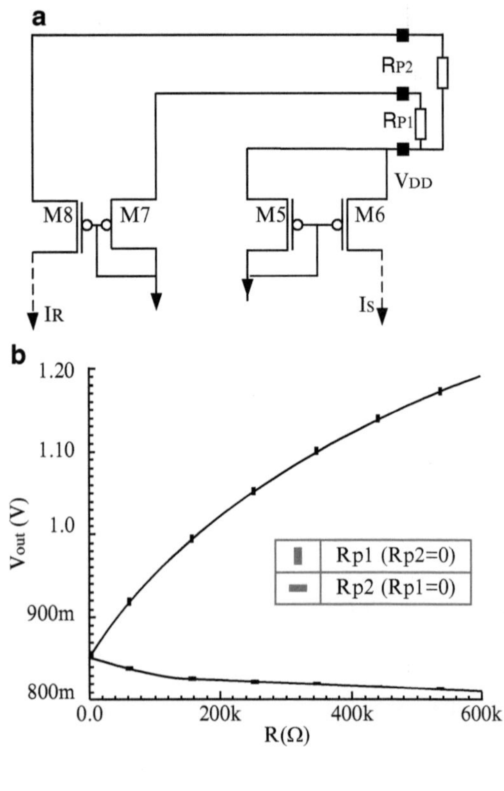

Fig. 4.27 Calibration technique: (**a**) Circuitry along with R_{p1} and R_{P2} and (**b**) simulation results of V_{out} variations versus Rp1 and Rp2

Adjustable Current Gain

An adjustable reference current I_R can be implemented in order to diminish $V_{off.}$ using a dedicated circuit topology. As shown in Fig. 4.28, corresponding to each digital input D1 – Dn(Đ) the reference current I_R is generated through $M6_{11}$–$M6_{1n}$ and $M6_{21}$–$M6_{2n}$ with the aspect ratios shown in the following equation [32].

$$\frac{W_3}{L_3}=2\frac{W_{61}}{L_{51}}=2^2\frac{W_{52}}{L_{62}}=\ldots=2^n\frac{W_{6n}}{L_{6n}} \tag{4.25}$$

where W_{6i} and L_{6i} are the width and length of transistors $M6_{1i}$ for i = 1 to n. Also, W_3 and L_3 are the width and length of transistor M5. The digital input data D_1... D_m(Đ) in this design can generated by programmable unit integrated either on-chip or through the off-chip platform. It should be mentioned that any residual offset even after applying an offset cancellation procedure is automatically eliminated by the subtractions of two subsequent measurements prior to and after analyte injection. Therefore V_{off} has no significant effect on the accuracy of capacitive detection and it only deteriorates the output dynamic range.

Further Modification

A wide swing current mirror (M_9–M_{12}) is used (Fig. 4.28) instead of simple current mirror shown in Fig. 4.21 (M_9–M_{10}). By applying the appropriate bias voltages V_a and V_b, a unity gain current mirror is implemented to conduct I_R toward the output node as described in this section.

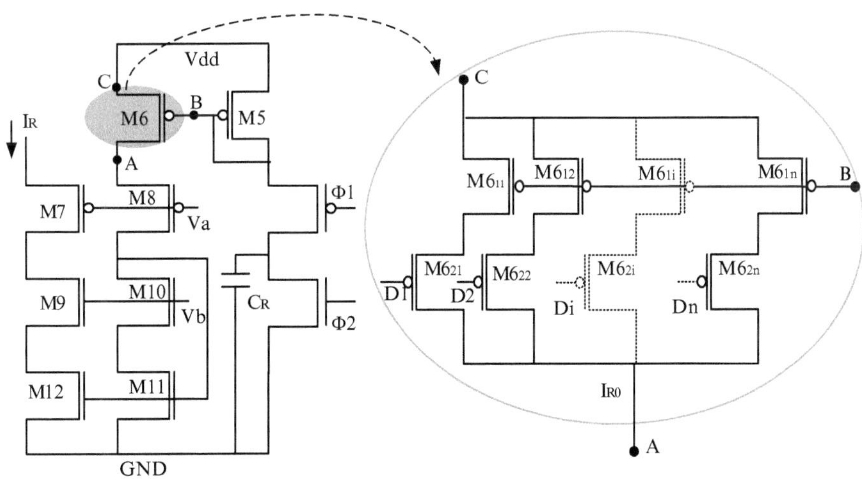

Fig. 4.28 Offset cancellation circuit featuring wide swing current mirror and variable gain current mirror

Practical Considerations

The viability of the circuit shown in Fig. 4.28 for LoC applications was demonstrated by applying chemical solvents. As shown in Fig. 4.29a, b, by applying pulse signals Φ1 and Φ2. As seen in these figures, V_{out} rises up ($f = 1$ kHz) when the interdigitated electrodes were exposed to dichloromethane, and methanol respectively and it decreased during the discharging period. As anticipated, the corresponding V_{out} for Methanol is higher than one for dichloromethane. This is because the dielectric constant of methanol is higher than the dielectric constant of dichloromethane. Figure 4.30a, b shows the sensor chip fabricated using CMOS process developed for this purpose. As shown in Fig. 4.30a, this chip features an interface circuit (Fig. 4.21) and two sensing and reference electrodes at both ends of the die. A microfluidic packaging method [270] was also performed to on this die to create a microchannel above the sensing electrodes (Fig. 4.30b) where the reference electrode is insulated by an optically transparent epoxy.

The same circuitry was employed to implement another chip with an array of sensing electrodes [269]. The experimental results presented in Fig. 4.31 show that once the analyte is introduced to sensing electrode, the corresponding voltage

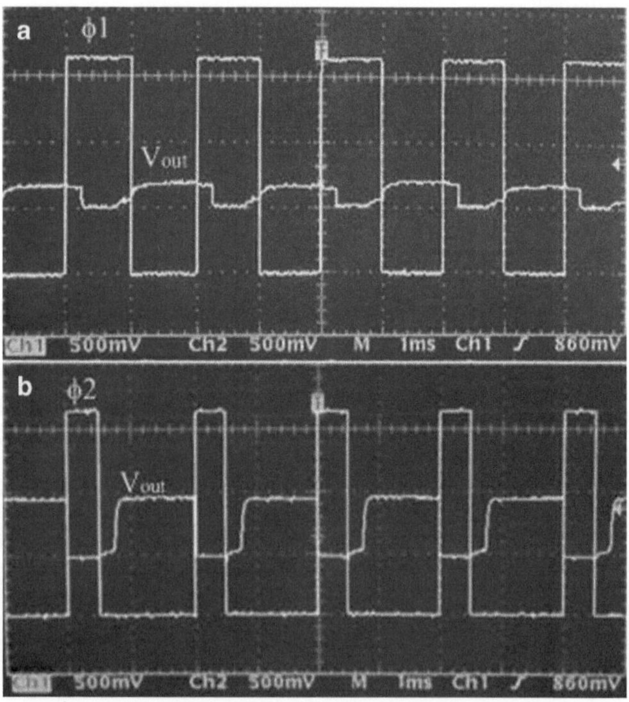

Fig. 4.29 Measurement results for differential current core-CBCM circuitry by applying (**a**) dichloromethane and (**b**) methanol [166]

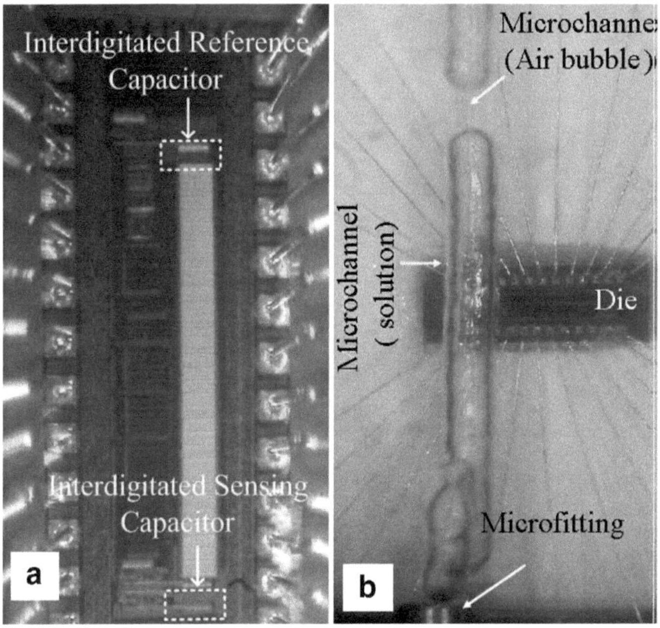

Fig. 4.30 Optomicrograph of fabricated circuitry shown in Fig. 4.21: (**a**) die and (**b**) microlfuidic packaging [206]

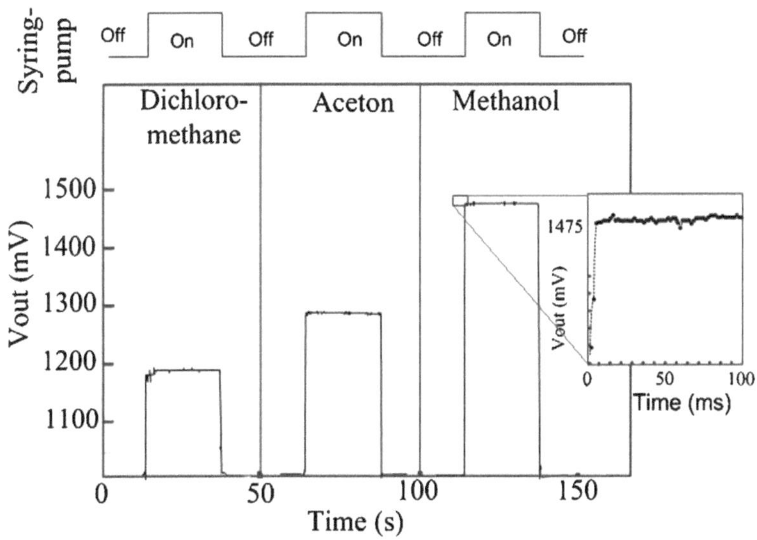

Fig. 4.31 Recorded V_{out} when the microchannel is empty (after calibration), or filled with dichloromethane, acetone and methanol [66]

appears in the output of the sensor as fast as the speed of the interface circuit. In this work, the sharp output variation of the sensor is due to direct measurement of the analyte without an extra sensing layer on top of the sensing electrodes. For example, using a polyetherurethane (PEUT) sensing layer for chemical detection in a gas-phase fluid causes a linear sensitivity for a wider dynamic range of analyte concentration and consequently a slower response of the sensor [73]. Therefore, with an additional sensing layer, the response time can be greater than the results shown in this figure.

It should be mentioned that a specific set-up is required for such measurements. As shown in Fig. 4.32, a stereo microscope can be used to observe the solution which is injected with a syringe pump into the inlet of the microfluidic channel. Simultaneously, the oscilloscope monitors the periodic V_{out}. In Fig. 4.32, the sensor chip can be exposed to hydrogel fibre for the characterization of the chip [236]. A microelectronic platform (FPGA SPARTAN-3, Xilinx [64]) should support the sensor chip for data recording and help the search for optimum measurement conditions (e.g. offset cancellation).

The analog signal was initially converted to digital values and then transferred to a computer. A fully integrated capacitive sensors includes an analog to digital converter (ADC) most often with 1-bit digital output. A sigma-delta modulator is a good architecture to readout the sensory signal.

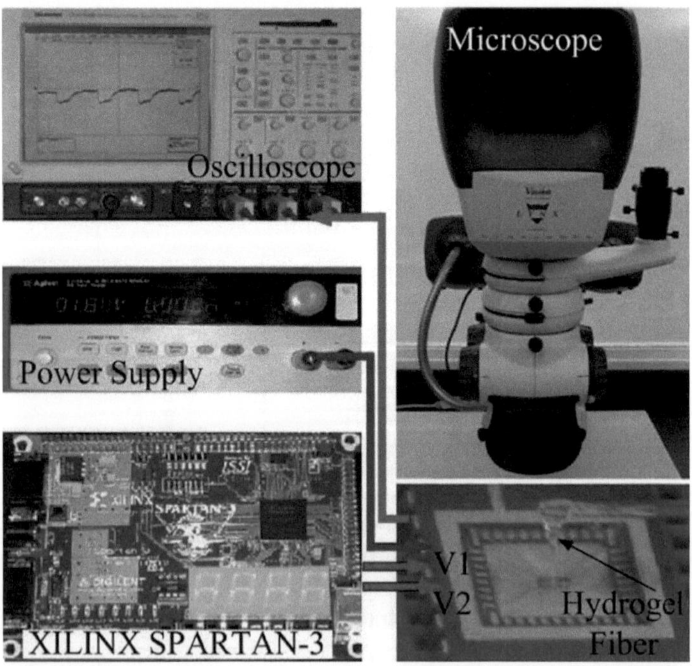

Fig. 4.32 Measurement set-up of capacitive sensor dedicated to LoC applications

4.4 Core-CBCM ΣΔ Capacitive Sensor

The presence rather than absence of a bioparticle in a LoC system results in a DC value which can be converted to digital data through a DC input ΣΔ ADC. Low complexity and low speed ΣΔ converters are good alternatives for these applications. In fact, ΣΔ modulators offer the advantages of simple architecture, high resolution, easy implementation, small size and low power consumption which have big impact on high sensitivity integrated sensor arrays.

4.4.1 Definitions

The DC input first-order ΣΔ modulator is shown in Fig. 4.33 where $q(y_n)$ is the quantized value, which is defined as follows.

$$q(y_n)=\begin{cases} +b & y_n \leq 0 \\ -b & y_n > 0 \end{cases} \tag{4.26}$$

where (−b,+b) is the full dynamic range, b is the quantification content and e_n is the error which is accumulated in the delay integrator [271].

In this first-order ΣΔ modulator , x_n and y_n represent the input and output of an integrator.

$$y_n = y_{n-1} + x_{n-1} - q_{n-1} \tag{4.27}$$

It should be mentioned that in a DC input ΣΔ modulator, the input value x_n is almost constant or the variations occur at a very low frequency.

4.4.2 Charge to Digital Converter

A ΣΔ readout circuit can be realized by incorporating a voltage comparator to the corresponding core-CBCM circuit. As shown in Fig. 4.34a the integrated voltage on C_{in} is compared with a reference voltage and consequently the output pulse of comparator (q_n) is applied to the switch sw1 in series with I_ξ. I_ξ is generated by adding a transistor (M_{6n+1}) to the circuit shown Fig. 4.34b.

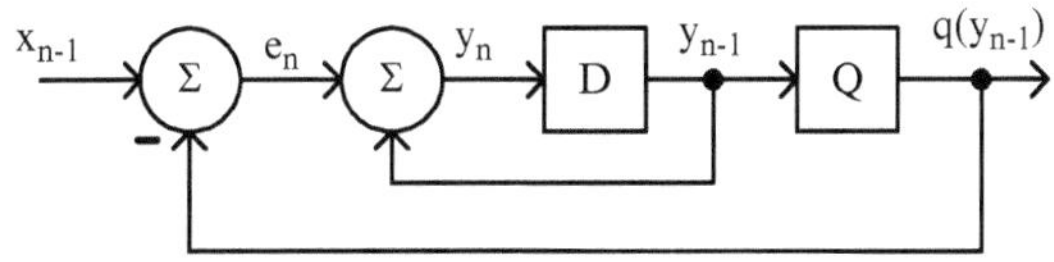

Fig. 4.33 Discreet time model of a first order ΣΔ modulator (Σ: summation, D – delay, Q – quanizer)

In fact, I_ξ instantaneously charges C_{in} resulting in a DC voltage change on V_{out}. After a few iterations corresponding to C_s, V_{out} can be larger than V_R so that sw1 is closed and V_{out} falls down to another level lower than V_R (see Fig. 4.34c). As shown in Fig. 4.35a, in each period of applied clock pulse, I_S is generated and subsequently the voltage on C_{in} is incremented up. In other words, the integrating C_{in} functions

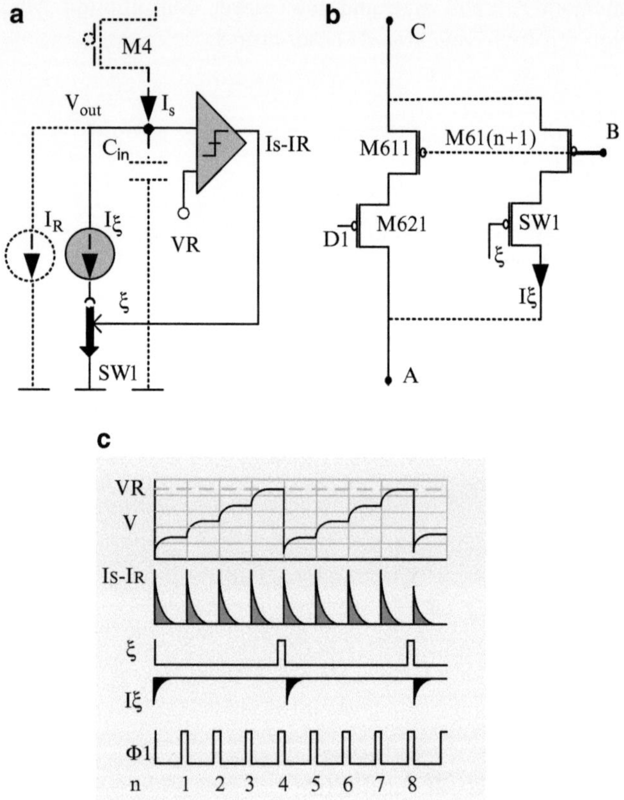

Fig. 4.34 Core-CBCM ΣΔ modulator: (**a**) circuit, (**b**) 1-bit DAC, and (**c**) pulse diagram

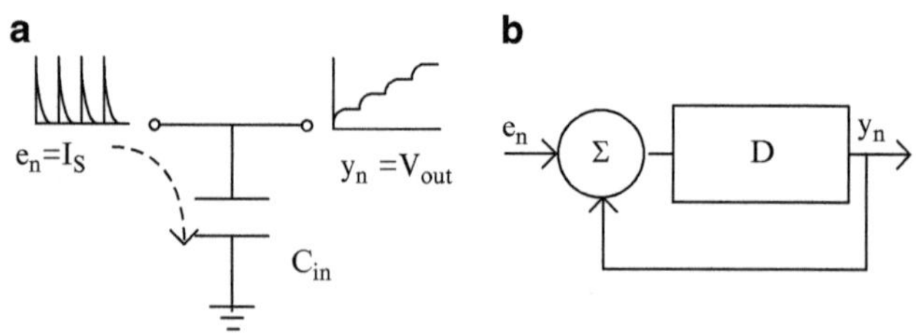

Fig. 4.35 Simple integrator in core-CBCM ΣΔ ADC: (**a**) circuit and (**b**) equivalent model

similar to a delay integrator as illustrated in Fig. 4.35b. As shown in this figure, the current error is converted to a voltage error and accumulated to the output of capacitor's output.

This current error is generated by the circuitry shown in Fig. 4.36. The current mirror (Q1, Q2) and sw_1 together constitute a 1-bit digital to analog converter (DAC) for this first order ΣΔ ADC, so that the digital input (ζ) is converted to I_ζ by this 1-bit DAC. This current is subtracted from $\Delta I = I_S - I_R$ resulting in e_n.

The quantization is also performed by the voltage comparator as shown in Fig. 4.37, a track and latch voltage comparator is followed by a RS flip-flop [272]. This comparator consists of a differential amplifier (M15-16) sensing the

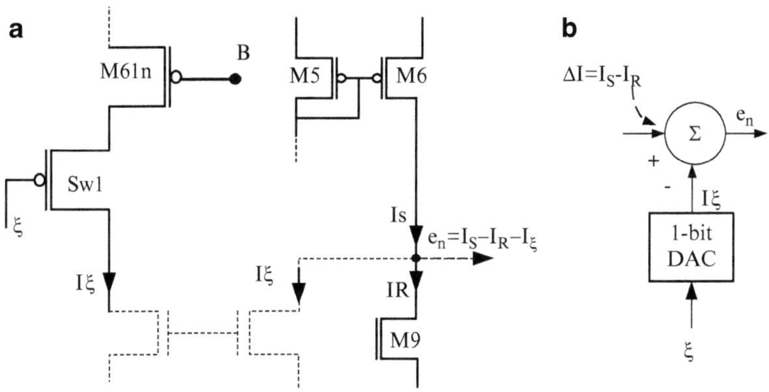

Fig. 4.36 One-bit DAC of core-CBCM ΣΔ ADC: (**a**) circuit and (**b**) equivalent model

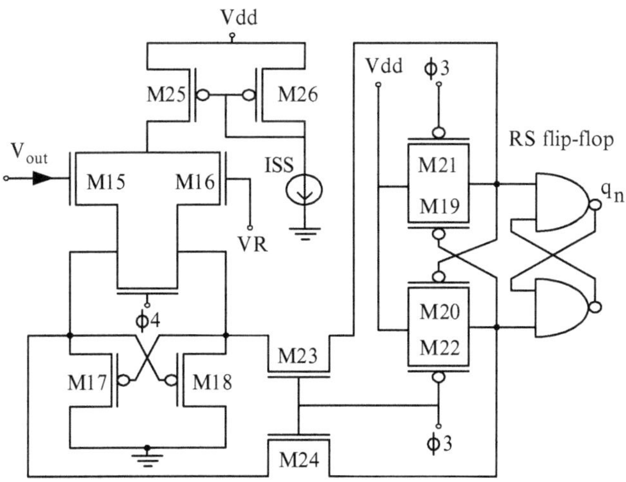

Fig. 4.37 A track and latch comparator

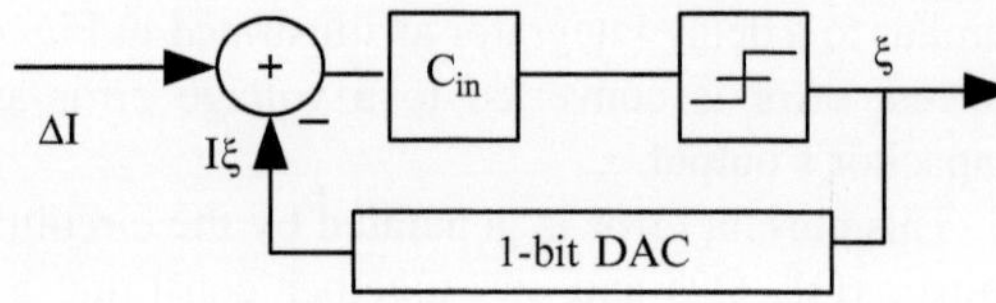

Fig. 4.38 Model of core-CBCM ΣΔ modulator

input differential voltage. The logic states of M19-20 results from the generated drain current on M15-16 where Φ3 is high and Φ4 is low. Once Φ3 becomes high, the regenerative process starts on M19-20 resulting from the initialization of M17-18 logic states. The RS flip-flop operates as a latch to save the logic states of M19-M20 during one clock pulse period.

Based on the above discussions, the core-CBCM ΣΔ modulator can be modeled in the diagram shown in Fig. 4.38. In this modulator, the input ΔI is a linear function of input sensing capacitance. Therefore, the digital output of the comparator (ζ) is proportional to C_S.

4.4.3 Discussions

The sample based relation of this modulator is shown in Eq. 4.28a for each period/sample of its operation which is comparable with the ideal relation of the DC input first order ΣΔ modulators stated in Eq. 4.28b [273].

$$y_n = y_{n-1} + x_n - q_{n-1} \tag{4.28a}$$

$$y_n = y_{n-1} + x_{n-1} - q_{n-1} \tag{4.28b}$$

where y_n, x_n and q_n are the nth sample of the integrator, input and output respectively. As the injected charge in C_{in} is instantaneously integrated x and y are simultaneously elevated. This is why both x and y in Eq. 4.28a have the same index n where in Eq. 4.28b, the index of x and y are not the same. The different indexes in above relationships only results in a difference between the four primary bits of the output sequences, so that the circuit shown in Fig. 4.34a can still be considered a new realization of the first order DC input ΣΔ ADC.

Let us assume a constant input value $k = x_n$ and then obtain q_n and y_n correspondingly for k = 0, ¼, ½, ¾ using this new ΣΔ modulator. By applying the input values, the relation of the new modulator ($q_{n-3} = Q_n$) can be established for $n > 3$ where Qn is the output sequence of an ideal first order, DC input ΣΔ with the same input (Fig. 4.39) [220, 273].

By applying a summation operation on both sides of $y_n = y_{n-1} + x_{n-1} - q_{n-1}$, we can obtain the following equation.

$$x_n = k = q_1 + q_2 + \ldots q_n)/N \tag{4.29}$$

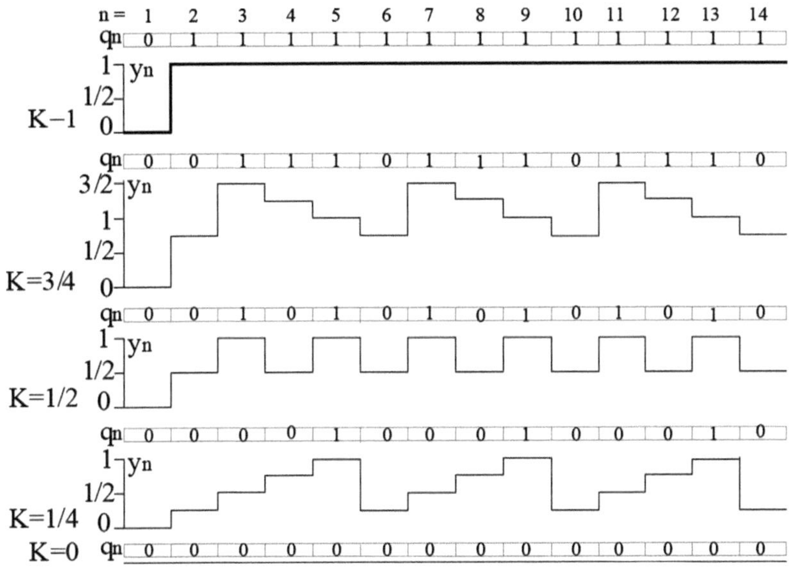

Fig. 4.39 Dicreete simulation results on core-CBCM ΣΔ ADC

where $n \geq 1$ and $y_N = 0$. By applying the same calculation to Eq. 4.28b the same results are obtained. Therefore, this simple calculation not only offers a simple technique to decode the data, but also proves the viability of this technique.

4.4.4 Circuit Level Simulation Results

As anticipated, the simulations in Fig. 4.40 show different pulse streams for two different values of ΔC. As seen in these simulations, there is a time delay for the generation of the first bit at the output of modulator. This time is inversely proportional to sensing capacitance. After this time, according to C_S, a 1-bit digital word is generated.

The discreet Fourier transform (DFT) of the output bit stream can also be obtained through simulation by SpecterS in Cadence as shown in Fig. 4.41. The DC component of this spectrum is equal to the mean value of the digital output signal and based on Eq. 4.29, this value is proportional to input value x.

Based on the above discussions, in order to validate that there is a unique bit-stream for each input ΔC, the corresponding input x was obtained using DFT technique as shown in Fig. 4.42 This figure reveals a linear relation between the output and input of the decoding module especially for the large number of bits ($Nc \gg 1$). In order to show that there is a unique bit-stream for each input DC, the average of each sequence (the number of one's per the number of pulses in each sequence for everything above the fourth bit) was obtained and shown in Fig. 4.41b. This figure reveals a linear relation between the output of the decoding module and ΔC.

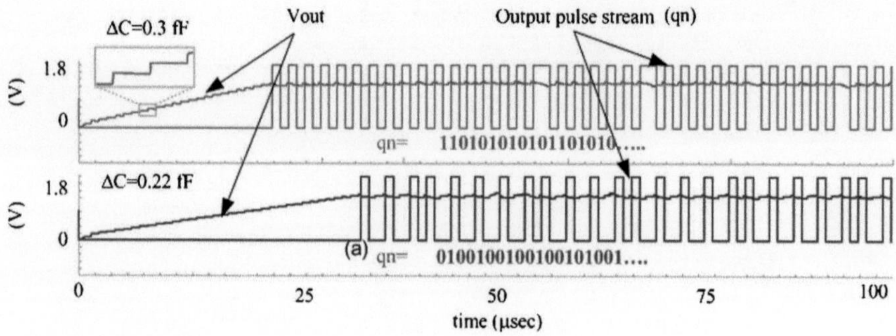

Fig. 4.40 Circuit level simulation results of core-CBCM ΣΔ ADC [66]

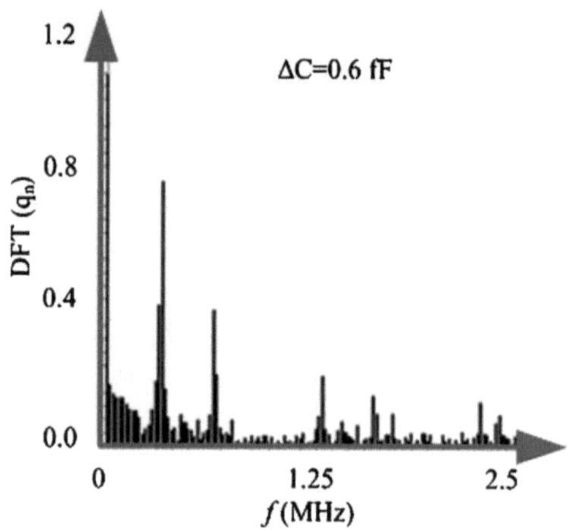

Fig. 4.41 DFT of output signal of core-CBCM ΣΔ ADC

4.4.5 Decoding Technique

In general, the output sequence of the first-order DC input ΣΔ modulators are periodic and the period (N) of sequence is dependent on the input value. For example, for an input $k > 0$, if $1/k$ is a natural number, this number will be the period of output sequence. Obviously, the Fourier series of such periodic sequences for each value of k is limited to the number of harmonics and is not similar to the noise-shaped spectra over-sampling ΣΔs [271] (see Fig. 4.41).

4.4.5.1 Simple Decoding Method

The information contained in the periodic sequences of such ΣΔ modulator can be extracted using the simple logical operations and it is not needed to employ the conventional decimation techniques using the finite impulse response (FIR) digital filters.

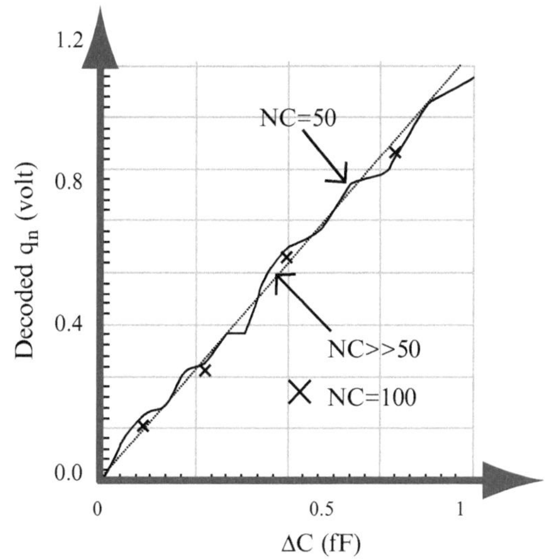

Fig. 4.42 DC component of output of core-CBCM ΣΔ ADC versus ΔC

By applying a summation operation on both sides of $y_n = y_{n-1} + x_{n-1} - q_{n-1}$, Eq. 4.30 is obtained.

$$\sum_{n=1}^{N} y_n = \sum_{n=1}^{N} y_{n-1} + \sum_{n=1}^{N} x_n - \sum_{n=1}^{N} q_{n-1} \tag{4.30}$$

where $n \geq 1$ and $y_N = 0$. Consequently by simplifying this relation, Eq. 4.29 can be obtained. The stream sequence q_n simply triggers a counter resulting in $q_1 + q_2 + \ldots q_N$ for a period N. Therefore, for a periodic sequence of a period N, the input x_n is equal to the output of this counter divided by N. For example, for a periodic sequence like the one shown in Eq. 4.31.

$$\underbrace{\overbrace{000\ldots01}^{M-1}\overbrace{000\ldots01}^{M-1}\ldots\overbrace{000\ldots01}^{M-1}}_{P}\overbrace{000\ldots01}^{M-2} \tag{4.31}$$

k is equal to (P + 1)/(P · M + M − 1). The noise contribution and the consequent accuracy of the decoded output are dependent on the number of limited cycles (or the number of bits in the sequence) N_C ($N_C \neq nN$, n J natural number) allowed for the conversion. The higher N_C, the lower ΣΔ noise contribution and consequently the higher resolution of the resulting ADC are expected. To date, several papers have reported the optimum decoding of the first-order DC input ΣΔ [36], and here we do not intend to elaborate on such issues. We will wrap up this discussion only by describing an optimum decoding technique for a core-CBCM ΣΔ modulator.

4.4.5.2 Optimum Decoding Methods

The challenge of decoding procedures is to recover the information from a 1-bit flow stream generated by a modulator rapidly. To date, several algorithms including the Zoomer [36], the Robust O(nlog(n)) for optimal decoding [274], and the rational cycle decoding algorithms (RCDA) [275, 276] have successfully been developed for this purpose. In all these algorithms, the output value is obtained through an iterative technique resulting in a long processing time. By combining such precision decoding algorithms with a dynamic procedure, it is expected a significant improvement in processing speed will result, which is critical in the design characteristics of readout systems especially for sensors arrays. Among these three decoding methods, in this section, RCDA is efficiently employed to develop a new algorithm suitable for the dedicated application [277].

Rational cycle decoding algorithm (RCDA) is a new iterative algorithm that improves Signal-to-Quantization Noise Ratio (SQNR) performance in comparison with other decoding algorithms [275] RCDA significantly modifies this important factors and is more robust than Zoomer regarding the disturbance factor [276]. RCDA can be applied on a sequence generated by a modulator in order to estimate the DC input of that modulator. By applying some modifications over the algorithm of RCDA, we have recently reported a RCDA [278]. The functionality of dynamic decoding architecture was demonstrated using AFS600 FPGA from Actel.

4.5 Core-CBCM Capacitive Sensing System

4.5.1 A System Level Realization

A core-CBCM capacitive sensor array is shown in Fig. 4.43. This system features an array of interdigitated electrodes which are used as the sensing electrodes C_{S1}–C_{S3}). These electrodes are realized atop the CMOS chip as already discussed in Chapter 2. Each sensing electrode is connected to a unit of interface circuit (CVC). An adjustable current mirror (see Fig. 4.28) is also derived by digital input data (D1, D2, …, Dm). This chip requires both analog (V_a and V_b) and digital (D_{1-m}) I/O signals. In this design, sensing electrodes are selected using digital addressing lines S_1, S_2 and S_3 ($\check{S}$) while, control logic includes reset and clock signals Φ1–Φ2.

Prior to analyte injection, the calibration procedure should be performed by finding the optimum value of digital input data D_{1-m}. For this, an off-chip module is commanded to start the sensor calibration before performing a measurement. This supporting module is realized in a FPGA platform that features a pulse generator, offset cancellation, decoding and sensor addressing units. The offset cancellation unit is used to make the decision criterion in the calibration algorithm realized by the off-chip module. In each period of clock Φ1, if V_{out} is larger than a threshold voltage (V_{th}), D_{1-m} is incremented until it reaches the desired value. Considering the aspect ratios of M_{C1}–M_{C8}, initially I_R is less than I_S, so the calibration process always starts with a value of V_{out} larger than V_{th}.

The output bit stream of the ΣΔ modulator is applied on the decoding unit in order to determine the sensing capacitance. This value is also used by the offset

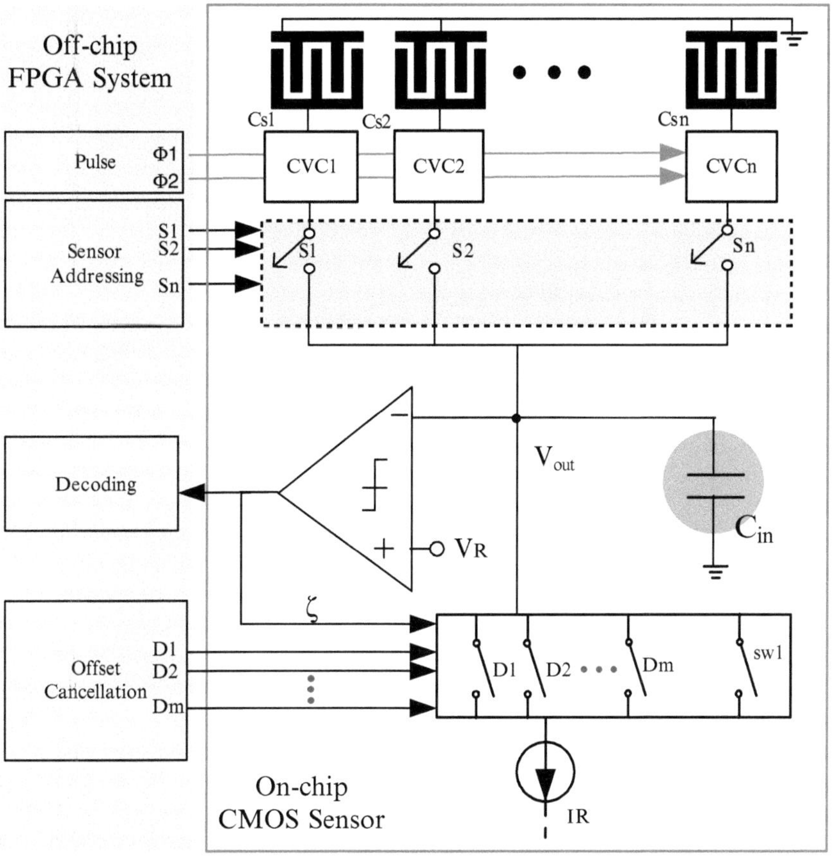

Fig. 4.43 A core-CBCM capacitive sensing system

cancellation (OC) unit. A FPGA platform (AFS600, Actel) was previously used to support the on-chip CMOS sensor [66].

4.5.2 *Experimental Procedures*

A core-CBCM CMOS sensor was reported. This system featured a reference electrode and three sensing electrodes. Figure 4.44a shows an optical microscopic image of an implemented sensor chip.

As seen in this figure, the sensing and reference capacitors were realized atop CMOS chip where the integrating capacitor was created in deep chip (Fig. 4.44b). The OC procedure was performed while the channel was empty. As seen in Table 4.1, the number of iteration and the residual voltage were different for different dies

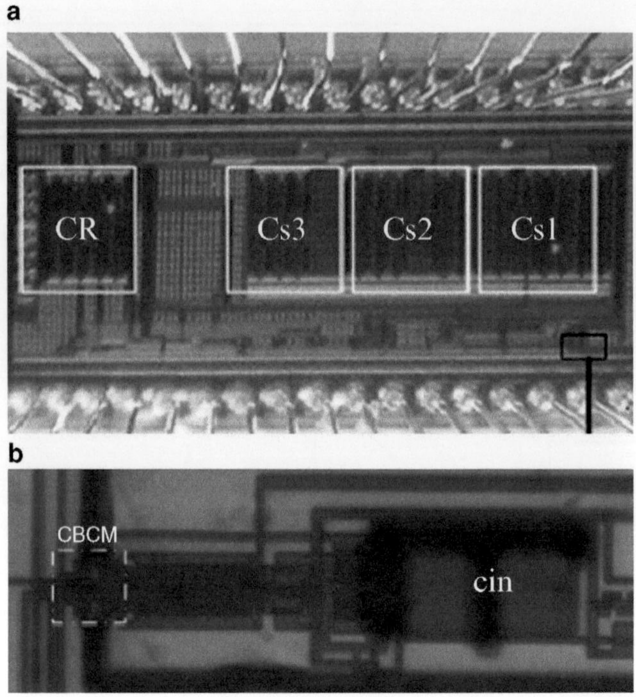

Fig. 4.44 Fabricated capacitive sensor chip: Optical microscopic image of (**a**) die and (**b**) the layout of interface circuit

Table 4.1 OC Procedure results on different dies (V_{th} = 535 mV)

Die #	Iteration #	V_{off} (mV)
1	26	0.551
2	89	0.527
3	37	0.538
4	24	0.549
5	100	0.530

Table 4.2 Decoded output of ΣΔ ADC for injected dichloromethane in channel

1	0	1	1	0	0	1	0	0
2	0	1	1	0	0	1	1	1
3	0	1	1	0	0	1	1	1
4	0	1	1	0	0	1	0	0
5	0	1	1	0	0	1	0	1
6	0	1	1	0	0	1	1	0
7	0	1	1	0	0	1	0	1

which could be due to mismatch error in the process and/or non-uniformity of the pad-etching. In the second mode, once again, dichloromethane was used to test the proposed system. Table 4.2 shows the decoded bit stream (q_n) at different times.

A cleaning procedure including hot water washing and air blowing along with temperature treatment was performed after each measurement [173].

As seen in this table, the first two bits of sensor's output are the representative of non-stable logic states. On other words, the resolution of the capacitive sensor is not better than 6 bits. Figure 4.45a shows the effect of dielectric variation on the sensor response. In this figure, the subtraction of the averaged measurements on the three sensing electrodes prior to and post solvent injection are shown in this figure. In another effort, two reference and sensing electrodes were employed for bacteria sensing purpose. Two microfluidic channels were implemented

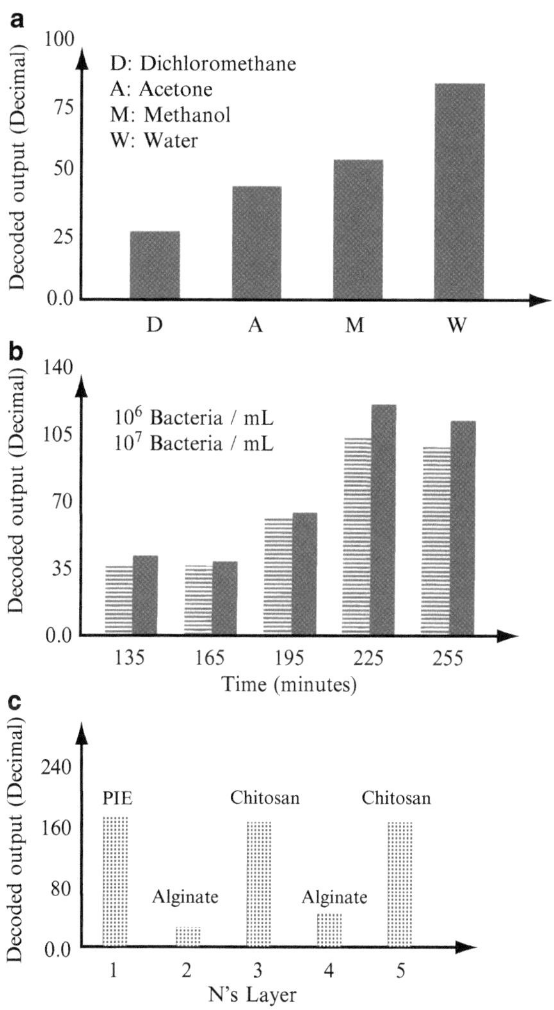

Fig. 4.45 Measurement results of core-CBCM capacitive sensing system for (**a**) organic solvent detection, (**b**) bacteria growth monitoring, and (**c**) polyelectrolyte layer detection

atop a CMOS chip. A bacteria solution including *E. coli* suspended in LB were injected in one channel while the second channel was filled only with LB. The bacteria growth monitoring was demonstrated through this sensing system as shown in Fig. 4.45b.

The formation of alternating chitosan and alginate layers was also performed on the top of a core-CBCM capacitive sensing system. As anticipated and seen in Fig. 4.45c, the top most conductive layer cause a high value in the recorded digital data ($V_{Max,i}$, i = first, third, fifth layer). The variation of $V_{Max,i}$ from one layer to another is relatively low in comparison with the average of $(V_{Max,1+} V_{Max,3+} V_{Max,5})/3$. On the other hand, as anticipated, this figure shows the low values of the recorded data corresponding to double layer formation (second and fourth layers). Based on the above experimental results, the formation of thin layers of positively and negatively charged layers can be monitored using the proposed capacitive sensor [220].

4.6 Summary

We reviewed the recent significant progresses of CMOS capacitive interface circuits including the SC interface circuit, time-constant sensing method, the capacitive inverter amplifier and core-CBCM methods dedicated for LoCs. The viability of these techniques for LoC applications were demonstrated and discussed by showing the appropriate biological and chemical testing results. As CMOS capacitive sensors as the result of offering low complexity techniques have received great attentions, it is required to design a generic capacitive sensing system which can be suitable for various applications. Further efforts should be made in the future to develop such a generic system with a large array of sensing electrodes that can be integrated with appropriate functionalized sensing layers for biological applications.

Chapter 5
Microfluidic Packaging Process

A CMOS-based LoC system would require efficient microfluidic packaging to protect the circuitry from the biological and chemical analytes, as well as the external environment. Microfluidic packaging is also critical to direct the fluids towards the embedded sensors or actuators for analysis. Ideally, these microfluidic packaging components, including micro-channels, -chambers, -fittings, -valves and -pumps should be performed using a low temperature process with reliable hermetic bonding [278]. The leakage of analytes (especially of charged molecules, as is the case with many bioanalytes) from microfluidic components may increase the parasitic capacitances or resistances and thus affect the circuit characteristics.

Conventionally, microfluidic packaging is performed using chemically-inert epoxy to cover the bonding wires or pads, or to underfill the flip-chips with an opening for free access of analytes. Since the thrust of current sensor development has been towards developing versatile platform technologies that could be adapted for a range of analytes, much of the current research centers on the incorporation of microchannels through MEMS processes onto CMOS chips [279–283]. Less attention has been paid on incorporating microfluidic components following chip fabrication procedures. However, standard, mutually-compatible platforms may allow CMOS-based LoC technologies to transition from a laboratory prototype to a ready-to-use product in the near future. With this in mind, to date, several microfluidic packaging techniques have already been reported that create microchannels atop a sample CMOS chip [284, 285]. In this chapter, a brief literature review of these techniques is put forward. Following this, the focus will shift towards the direct-write microfabrication process.

5.1 Microfluidic Packaging Methods

On-chip micromachining procedures, adhesive techniques, rapid prototyping techniques and direct-write assembly are described in this section.

E. Ghafar-Zadeh and M. Sawan, *CMOS Capacitive Sensors for Lab-on-Chip Applications: A Multidisciplinary Approach,* Analog Circuits and Signal Processing,
DOI 10.1007/978-90-481-3727-5_5,

5.1.1 *On-Chip Micromachining Procedures*

Surface micromachining is one of several conventional micromachining techniques that have been reported for on-chip microfluidic fabrication purposes. In this technique, a conformably deposited material is formed over a sacrificial layer, which is subsequently removed by etching to yield the microstructure or microchannel, as shown in Fig. 5.1a–c. Using this technique, Mastrangelo et al. reported the fabrication of plastic capillaries ranging from 0.5 to100 μm on standard CMOS chip [286, 287].

Another effort for on-chip microfluidic fabrication was performed by Rasmussen et al. without the use addition material deposition [288, 289]. A shallow microchannel was realized using the standard metal layer (aluminum) inside the CMOS chip. In fact, by using traditional CAD tools, a metal layer was selected and patterned on the top of the sensing site, as shown schematically in Fig. 5.2a. This conductor (and vias) played the role of the sacrificial layer, which was etched using 80% phosphoric acid, 5% nitric acid, 5% acetic acid, and 10% water (see Fig. 5.2b). This procedure was successfully employed to create a monolithic integration into a microelectronic interface circuit for sensing the flow rate of liquid [284]. A deep microchannel can also be fabricated through opening

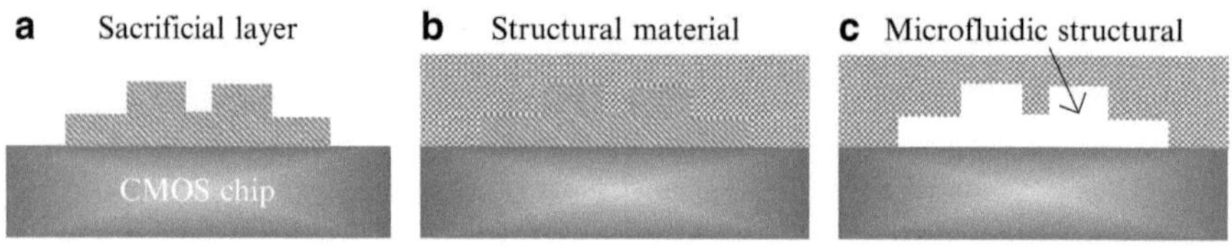

Fig. 5.1 On-chip micromachining procedure: (**a**) sacrificial layer deposition, (**b**) structural material deposition and (**c**) sacrificial layer removal

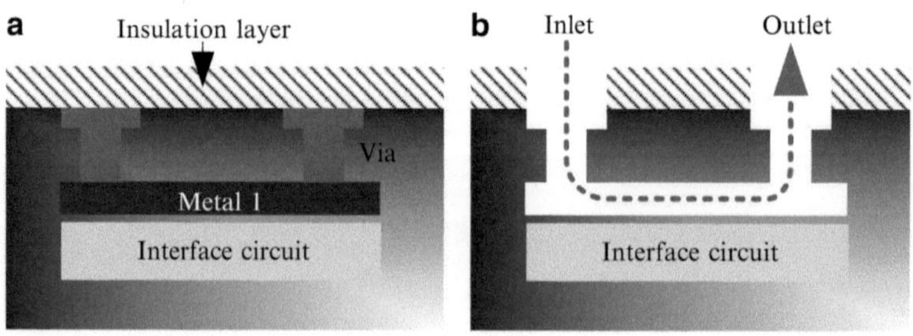

Fig. 5.2 Microchannel realization through CMOS process (**a**) before etching and (**b**) after etching

windows on top of chip and by using a different etchant, along with the multi metal layers as reported in [284].

Recently, Lee et al. proposed an IC/Microfluidic hybrid microsystem for 2D magnetic manipulation of individual biological cells [290, 291]. A spin-coated and patterned polyimide was formed on the SiGe IC as the sidewalls of a microfluidic channel. A glass cover slip was sealed on top of the sidewalls and connected to the inlet and outlet fluidic tube in order to circulate the biological solutions in the microfluidic system. This low temperature method could similarly be applied to CMOS sensors for other LoC applications.

Advantages and disadvantages: The microfluidic components can be fabricated using this technique with almost the same precision of standard CMOS technology in order to create a sacrificial layer of embedded conductors, without the need for extra deposition procedures, in clean room. However, as on-chip micromachining procedures are labour intensive and expensive, less attention has been paid to these procedures in comparison with other microfluidic packaging methods.

5.1.2 *Adhesive Methods*

A microfluidic structure can be fabricated using a variety of polymeric techniques (e.g. hot embossing [278, 292]), and adhesively bonded onto CMOS chips using glue or low temperature plasma bonding method (see Fig. 5.3) [292]. It is obvious that by using other bonding techniques with high temperature and/or high voltages, no electronic components can be present on the substrate, thereafter an adhesive low temperature technique can assure a reliable bonding method without any damage on underneath devices. Polymer based materials are widely employed using replication techniques. The two most common replication techniques are moulding [293] and hot embossing. As shown in Fig. 5.3a, b, a microfabricated mould can be used to pattern polymeric materials, such as PDMS. By separating the mould, a microstructure can be created (Fig. 5.3c) that can be affixed above the CMOS chip (Fig. 5.3d) Several epoxy microfluidic devices can be implemented by using these techniques.

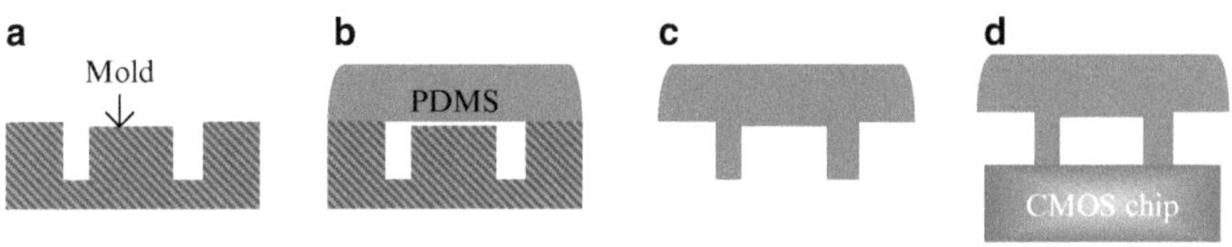

Fig. 5.3 Adhesive method: (**a**) microfabricated mold, (**b**) injecting PDMS, (**c**) PDMS microstructure, and (**d**) PDMS microstructure affixed on to a CMOS chip

Two examples are a nozzle-diffuser pump and a temperature control system for LoCs [294].

The hot embossing technique allows high precision replication of features from a mould onto thermoplastic materials. Chartier et al. successfully achieved the fabrication of a polymer-based microfluidic structure through hot embossing [292]. This was integrated onto a CMOS-based LoC for bioparticle detection and manipulation. In addition, a follow-up paper describes the fabrication of microfluidic networks on the same CMOS-based system using a dry film resist [295].

5.1.2.1 Advantages and Disadvantages

The main advantage of this technique is the creation of polymeric structures with very high precision that are very suitable for biological LoC applications. However, a problem associated with this technique is that adhesive methods cannot guarantee a hermetic bond for different solutions flowing through the microfluidic structure if fabricated on a rough-surface CMOS chip.

5.1.3 Rapid Prototyping Techniques

One important reason for fabricating or bonding a microfluidic structure on top of an IC is to cover the chip with an insulation layer, with an opening to access the sensing/actuating electrodes. Applying an epoxy-based packaging could be a fast and simple way of accomplishing this task [96]. A programmable dispensing system (e.g. Champion 8300 dispenser) is often employed to extrude the epoxy from a nozzle onto predetermined points on the sensor chip. Also, by using flip-chip packaging techniques (Fig. 5.4a) [96], it could also be possible to cover the sensor chip with another covering chip. Beside these methods, many other rapid prototyping methods can be performed to cover the chip with a window through which the analyte can be inserted onto the sensor chip. Among these techniques, Medoro et al.

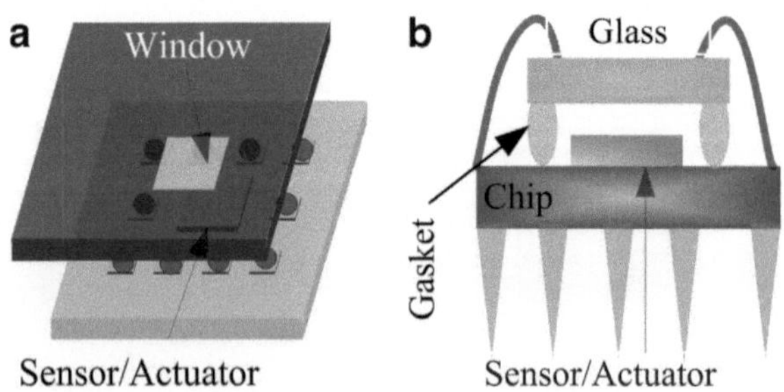

Fig. 5.4 Mocrofluidic packaging using (**a**) standard flip chip technique and (**b**) a rapid prototyping method

reported a non-conventional method using simple laboratory devices (see Fig. 5.4b) to create a well on top of the dielectrophoresis electrodes for cell manipulation [76].

Advantages and disadvantages: Even though such rapid prototyping methods can be performed at a very low cost in a laboratory environment, the rapid prototyping techniques cannot be applied for precision applications with a large number of wells and microchannels.

5.2 Direct-Write Microfabrication Process

In addition to the above-mentioned microfluidic packaging methods, direct-write microfluidic hermetically bonded packaging can efficiently be applied to CMOS based Lab-on-Chip. Herein, the principle of the direct-write technique is discussed, along with other practical considerations, advantages and associated problems.

5.2.1 Direct-Ink Writing

Based on recent literature, several direct ink writing techniques, such as ink-jet printing [296], hot-melt printing [297] and micropen-writing [298], are capable of forming materials in three dimensions. Among these techniques, this paper focuses on the Direct-Write Fabrication Process (DWFP), which is a filamentary-based method to extrude a paste-like material from a nozzle and deposit it onto a substrate. To date, a broad diversity of inks, including ceramic inks [299], have been designed for DWFP. A list of functional inks and their proposed applications is given in Table 5.1. Among these, a concentrated polyamine-rich ink, compromised of appropriate amounts of cationic and anionic polyelectrolytes, is used for biomimic applications [300].

Table 5.1 Designed Inks for DWFP and their applications

Ink	Microfabricated device	Application	Year	Ref.
Polyelectrolyte	3D Microscaffold	Biomimic of diatom frustules	2006	[300]
Polyelectrolyte	3D Si hollow-woodpile structures	Photonic Crystals	2006	[301]
Hydroxyapatite	Bone scaffolds	In vivo bone response	2007	[302]
Hydroxyapatite	Bone scaffolds	Repair of defects in bone	2007	[303]
Wax	3D Mixer	LoC	2003	[304]
Wax	Microfluidic packaging	LoC	2009	[305]
Wax	Optical waveguide	Biochemical sensor	2008	[306]
NIO and YSZ[a]	Interdigitized electrode	Fuel cell	2008	[307, 309]

[a](ZrO_2 doped with 8 mol% Y_2O_3) and $(La_{0.8}Sr_{0.2})_{0.98}MnO_3$-YSZ

A chemical vapor deposition procedure can be performed on this polyelectrolyte micro-scaffold in order to create 3D silicon hollow-woodpile structures for photonic applications [301]. In another effort, bio-compatible Hydroxyapatite (HA) scaffolds were employed using a DWFP process to investigate bone response [302, 303]. A fugitive ink composed of petroleum jelly and microcrystalline was developed as a sacrificial layer in the implementation of microfluidic structures [304]. In this chapter, we will describe the applicability of this ink for microfluidic technology by demonstrating the experimental results of several microfluidic devices, as mentioned in Table 5.1 [305–307]. In addition, the viability of DWFP to realize single-chamber, micro-sized solid oxide fuel cell (SC-μSOFC), with coplanar microelectrodes, is demonstrated for fuel cell applications [308, 309]. Among many different investigated techniques [310–313], the DWFP method has been shown to be a suitable approach for the fabrication of these cells [308–310]. Powders of the respective electrode material are simply processed to form inks, which are then extruded through micronozzles and robotically deposited onto the electrolyte substrate. Sintering of the deposited electrodes completes the fabrication process by consolidating the electrodes and creating a porous microstructure.

The main advantage of the DWFP technique for SC-μSOFCs with coplanar electrodes lies in the simplicity of the technique, as different electrode patterns – ranging from simple parallel electrode pairs to interdigitated and comb-like electrodes or even arbitrary, non-conventional designs [307] can rapidly be generated without the use for masks or molds. In addition, the use of viscoelastic, gel-like inks enables the creation of thick, three-dimensional electrodes. With the possibility of quickly creating electrode structures of different sizes, shapes and interelectrode distances, DWFP appears to be an appropriate tool for the further optimization of SC-μSOFCs with coplanar electrodes.

5.2.2 *Fundamentals of DWFP*

Figure 5.5 shows the basic functionality of this technique. By applying pressurized air, functional materials, or so-called inks, are deposited in a trajectory that is preloaded into the robotic system.

Several parameters, such as the speed of the robot (V), inner diameter of the nozzle (D_z), distance between the substrate and nozzle (H), air pressure (P) and ambient conditions, influence the performance of this process. The robot is programmed, using custom software, to carry a barrel of ink in a desired trajectory. The diameter of the ink filament (D_f) differs from D_z, and most often $D_f > D_z$. In order to deposit a fine ink filament (see Fig. 5.6a), the flow rate of extruded ink should be proportional to the speed of the nozzle and optimum values of V, P and H (V_0, P_0 and H_0) should be obtained precisely. Of course, the flow rate of ink is dependent on several parameters, including the length, diameter and surface roughness of the nozzle, the surface properties of the substrate and the ambient temperature. As shown in Fig. 5.6b, c, if the speed of the nozzle V_1 is less than V_0 (or the air pressure $P_1 > P_0$), the ink filament

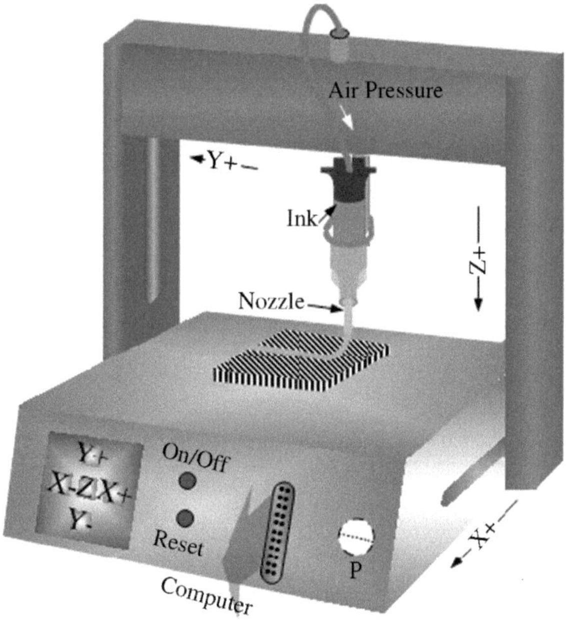

Fig. 5.5 Illustration of robotic based system for DWFP

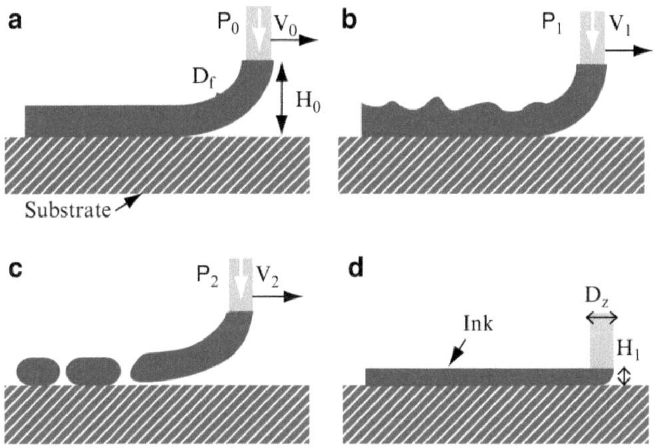

Fig. 5.6 Ink deposition on a substrate: (**a**) fine deposition, (**b**) deposition while $V_0 > V_1$ (or $P_1 > P_0$), (**c**) $V_0 < V_2$ (or $P_0 > P_1$) and (**d**) $H_0 > H_1$

will be deposited non-uniformly; if the speed of the nozzle V_2 is greater than V_0 (or air pressure $P_1 < P_0$), this will result in a discontinuous ink deposition. Additionally, if the height of the nozzle above the substrate is $H_1 < H_0$, the thickness of the ink filament becomes equal to H_0, which is less than D_f, as shown in Fig. 5.6d.

5.2.3 *Direct-Write Microfluidic Fabrication Process*

5.2.3.1 Procedures

Direct-write microfluidic fabrication process is performed using following three steps:

1. Sacrificial paste-like ink deposition through a micronozzle (Fig. 5.7a)
2. Encapsulation of ink filaments with liquid epoxy (Fig. 5.7b)
3. Ink extraction after the hardening of epoxy [304, 305] (Fig. 5.7c)

5.2.3.2 Fabrication Set-Up

The fabrication process is performed using a very low-complexity set-up consisting of a three-axis robot along with a dispensing device to extrude the ink architecture onto pre-determined sites on the substrate (e.g., silicon wafer, microelectronic die or PCB) as shown in Fig. 5.8. A robotic apparatus (Model I&J 2200, I&J FISNAR Inc.) controlled by a computer (JR Points, I&J FISNAR, Inc.) is used to perform the deposition pattern. The air-operated dispensing system (Model 2400, EFD, Inc.) can be used to extrude the fugitive binary organic ink [314] through different micro-nozzles (e.g. 10, 100 or 200 μm inner diameters).

The ink is either deposited onto a CMOS chip or some other substrate. V (≈1 mm/s) and H (≈250 μm for $D_Z = 200$ μm) and P should be kept constant during deposition. The values of these parameters were experimentally determined. The optically clear epoxy resin (Epoxide 835, Epoxitech Inc.) is poured to encapsulate the deposited ink filaments and cured at room temperature. Then, the part is heated to ~75°C in order to melt the ink, while air pressure (P ≈ 100 kpa) was applied at the inlet of the microfluidic channel for the ink removal. Fugitive ink is obtained from the combination of petroleum jelly and 25% microcrystalline wax. It should be mentioned that by increasing the percentage of microcrystalline particles in the ink, the viscosity of fugitive ink is increased and therefore a higher pressure is required in order to extrude it.

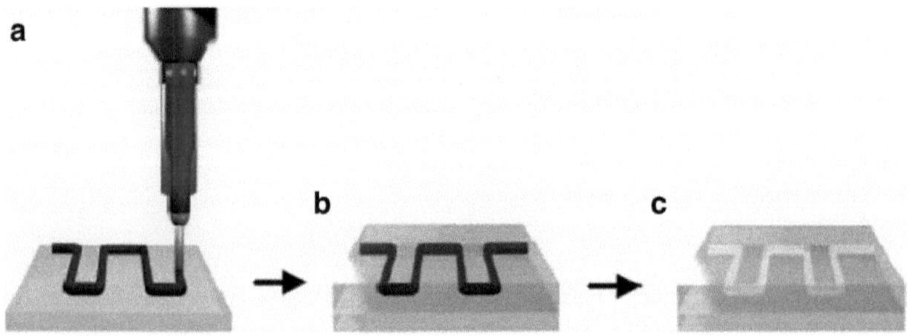

Fig. 5.7 Direct-write fabrication process: (**a**) ink deposition, (**b**) epoxy encapsulation and (**c**) ink removal

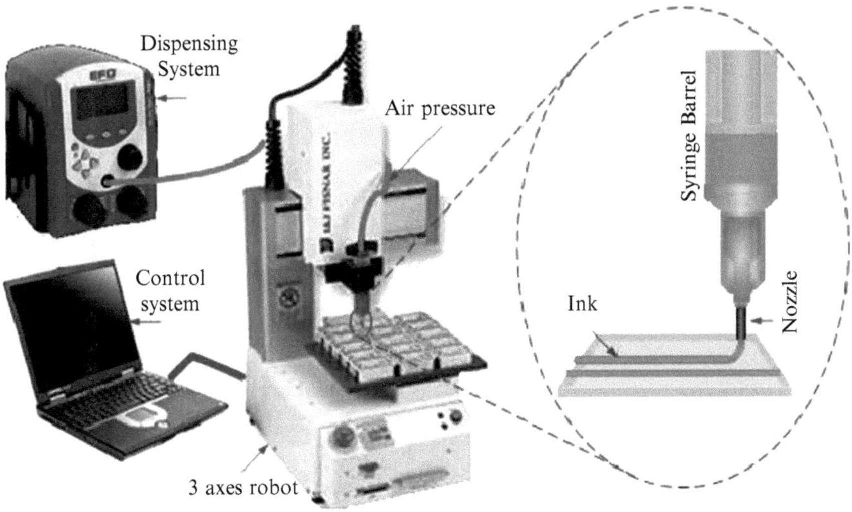

Fig. 5.8 A fabrication set-up for direct-write technique

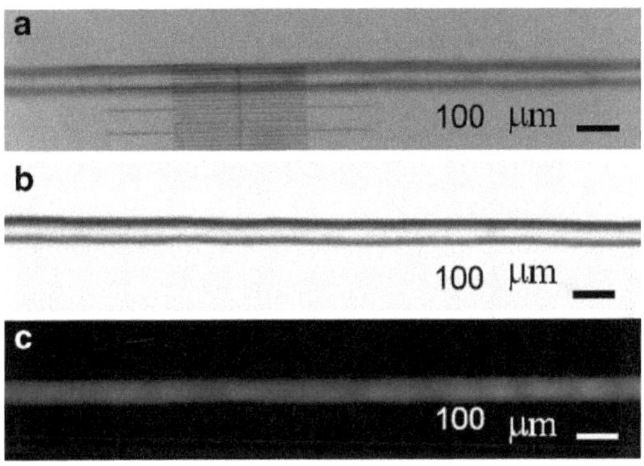

Fig. 5.9 Microfluidic fabrication process using DWFP: (**a**) ink filament on the top of glass substrate, (**b**) ink filament encapsulated with epoxy and (**c**) microchannel filled with a fluorescent dye

As shown in Fig. 5.9, the ink filament, encapsulated ink and channel all have the same dimensions. In fact, the ink filament is similar to the cross-section of the nozzle, and is capable of preserving its shape after epoxy encapsulation.

5.2.3.3 Advantages and Disadvantages of DWFP

DWFP is a low-temperature, low-cost process that can be employed to implement microfluidics with three dimensional and complex geometries using a very low complexity fabrication set-up. The required set-up system and materials for DWFP are not expensive and there is no need to deal with laborious and complex infrastructure relevant to microfabrication clean rooms. For this, DWFP can be a low complexity technique to implement the required microfluidic structures for chemical or biological researchers with less background in engineering.

Another advantage of DWFP is that microfluidics fabrication, integration and fitting connection can be simultaneously performed on any substrate, including microelectronic loose dies or chips. In a LoC system, there are two kinds of packaging: electrical and fluidic. Electrical packaging can be performed using conventional techniques, such as wire bonding or flip chip. Thereafter, DWFP provides the microfluidic components and inlet/outlet fittings. In most of microelectronic packaging labs, programmable dispensing systems (e.g. Champion 8300 dispenser) are one of the most important conventional packaging facilities already available. DWFP could be performed using such programmable dispensing systems since they offer similar functionality [305]. The precision of DWFP is not comparable to lithography and the minimum size of the microchannels implemented through DWFP is above 30 μm, which is larger than other microfluidic fabrication processes with nanometer depth [315]. For this, DWFP can effectively be employed for the implementation of three-dimensional structures and microfluidic packaging purposes that do not require very small dimensions.

A helical, cylindrical or conic microchamber can be implemented through DWFP, as shown in Fig. 5.10. This figure displays a microscopic image (Dynoscope, Lynex, Vision Engineering) of a helical microchannel filled with a coloured liquid is depicted in Fig. 5.10. For this, a filament ink has been deposited on a rotating micronozzle. The encapsulation and ink removal processes are performed as discussed in Section 5.2. As shown in Fig. 5.11, the rotating speed ω and displacement speed v play key roles in this process. Helical microchannels can be used as microscale mixers, centrifuges and fractionation or counter-current chromatography devices for biochemical sample processing and analyses.

An optical microscope image of a conic microfluidic chamber is shown in Fig. 5.12a. The illustrations of ink deposition trajectory, stacked layers and an optical

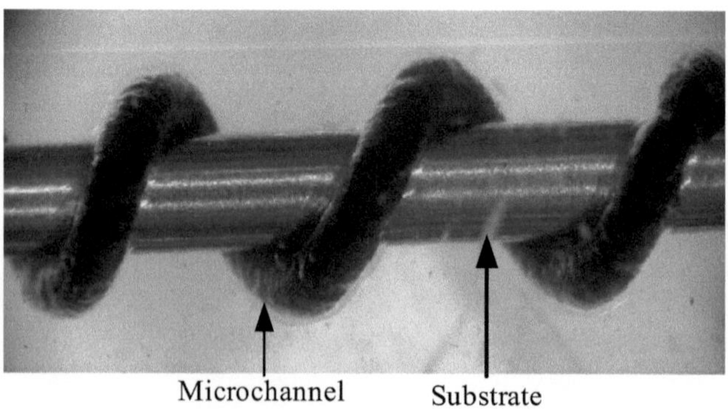

Fig. 5.10 Optical microscope image of helical microchannel filled with a coloured liquid [306]

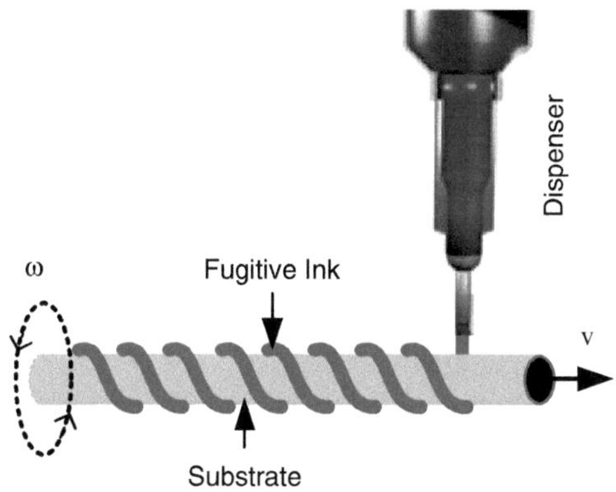

Fig. 5.11 Schematic of deposition set-up for helical microchannel (scale bar, 100 μm)

microscopic image of the corresponding ink architecture of this conic microfluidic chamber are shown in Fig. 5.12b–d, respectively. Also included in this figure is the UV microscopic image of an implemented microfluidic chamber with the same geometry of Fig. 5.12d but less volume of epoxy (Fig. 5.12e).

In another effort, filled-in cylindrical ink architecture was implemented, and its optical microscopic image is shown in Fig. 5.12f. In addition to the above mentioned advantages, this technique can be used for the water cooling of chips and printed circuit boards.

Figures 5.13a, b show an ink architecture and corresponding cylindrical microfluidic device which is filled with a fluorescent dye. For the implementation of this microfluidic device, the robot was programmed to pursue a helical path while the dispensing system extruded the fugitive ink.

5.2.3.4 Other Practical Considerations of DWFP

There are several practical issues which affect the performance of DWFP. Herein some of these issues are described.

Dispensing Model

DWFP can be easily adapted to conventional dispensing processes for fluidic packaging of LoC integrated sensors. To date, the dispensing processes of several fluids have been studied [316, 317] but the modeling of fugitive ink dispensing process is still in its infancy. With appropriate rheological models of the ink, liquid epoxy and pre-gel solutions, the complex microfluidic components could be optimized and

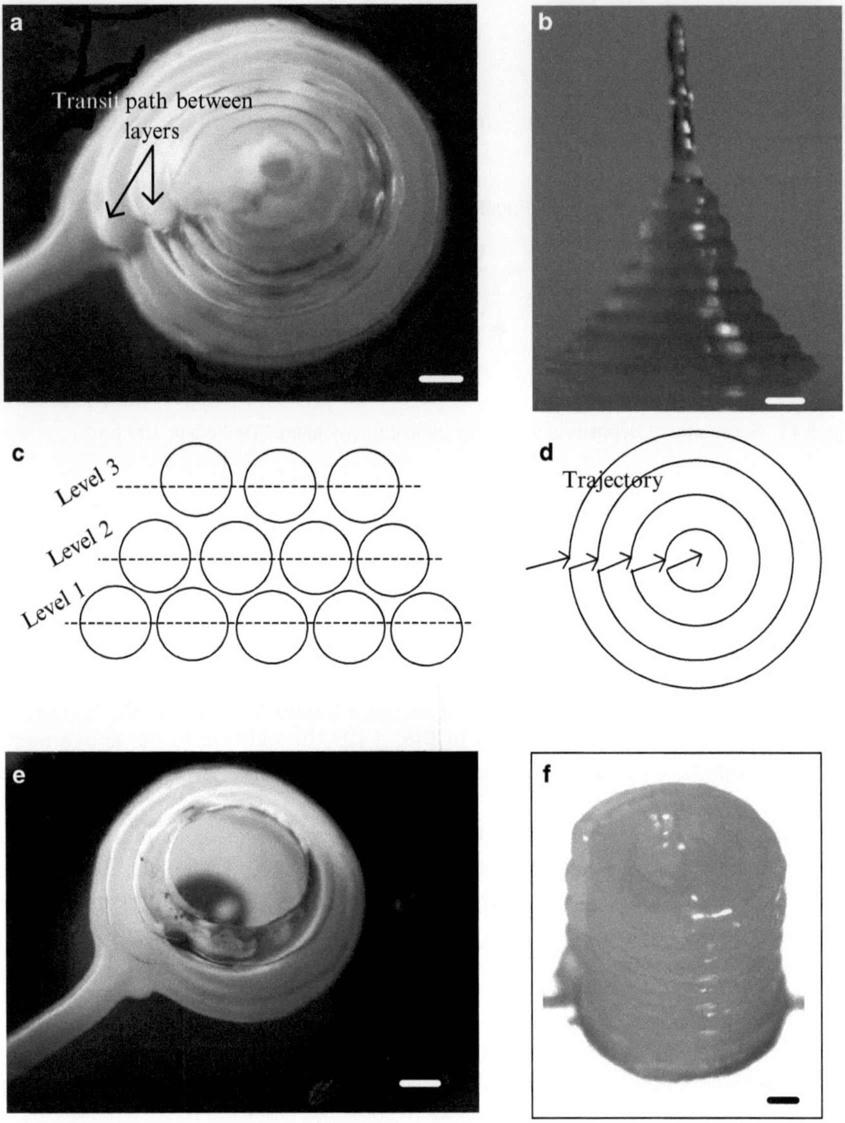

Fig. 5.12 Conic and cylindrical microreservoir: (**a**) conic microfluidic chamber, (**b**) conic ink architecture, (**c**) stacked layers (**d**) trajectory of ink deposition, (**e**) conic microfluidic chamber with less epoxy and (**f**) filled in cylindrical ink architecture (scale bare 200 μm) (scale 200 μm) [141]

implemented through an automated and programmable DWFP. Another approach to search for optimum values of V, H and P during the ink deposition is through neural networks (NN) [206]. In fact, the experimental data for different nozzle materials and lengths, inks and substrates, as well as different values of P, H and V, could be used to generate a NN model.

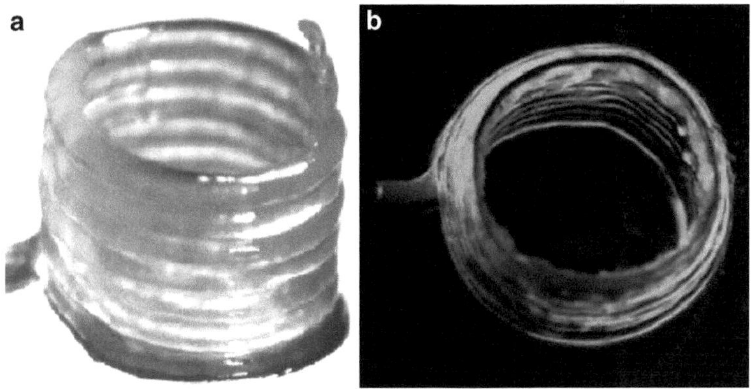

Fig. 5.13 Optical microscope image of (**a**) cylindrical ink architecture and (**b**) corresponding microchamber filled with fluorescent dye (outer diameter of cylinder, 2 mm) [141]

Micro-nozzle

The minimum radius of the channels fabricated by DWFP is dependent on the inner diameter of the micronozzle. The size of commercially available micronozzles, with metal tips (EFD Inc.), is around 100 μm, and with glass tips (World Precision Instruments, Inc.) is of the order of 100 nm. However, smaller glass tips are fragile and tend to break under high extruding pressure or as a result of accidental contact with the substrate surface. The fabrication of nozzles with different shapes and cross-sections for dispensing processes has already been reported [318]. Of course, by using nozzles with different geometries and cross sections, it could be possible to implement different microchannels. This can be an advantage when making large fluidic connections to the channel. Also, the implementation of smaller channels in proximity of sensing sites offers the advantage of low sample consumption and high sensitivity. The minimum inner diameter of commercially available nozzles fabricated in metal and glass is 100 μm (EFD Inc.)

Micofluidic Fitting

DWFP allows the simultaneous fabrication of microfluidic connections to standard fittings. Figure 5.14a shows the first step of fluidic connection by DWFP: the installation of the fitting close to the ink filament. The extruded ink from the fitting creates the connection to the microchannel (Fig. 5.14b). Then, the connection site, along with ink filaments, is covered with epoxy. The existing fugitive ink inside the fitting can be easily removed during the ink removal process. It should be mentioned that the integrated circuit is initially covered by an epoxy and then the ink deposition is performed. This is shown in Fig. 5.14c and it will be described in Section 5.5.

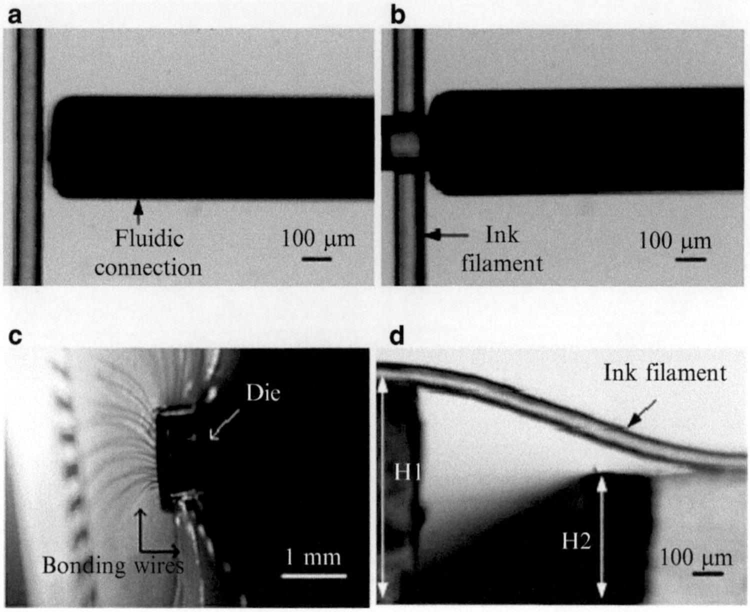

Fig. 5.14 Optical microscope images of (**a**) the fugitive ink and fitting, (**b**) deposited ink from fitting, (**c**) ink deposition between two substrates of different heights, and (**d**) epoxy resin surrounding the die [141]

Substrate for DWFP

DWFP can be performed on a wide variety of substrates (e.g. different materials and physical surface properties). Figure 5.14d shows an optical microscope side view image of a deposited ink filament between two substrates of different heights. This image demonstrates the flexibility of DWFP in the fluidic packaging process with different heights in the trajectory of deposition. This flexibility is a very important advantage of DWFP, as is apparent from Fig. 5.15. This figure shows a microchannel fabricated on a printed circuit board. Several microchannels can be performed on a PCB or chip to play the role of a cooling layer.

Microfluidic Structure

Epoxy is a good candidate for DWFP because it becomes very hard with no deformation during the ink removal process. If a soft bake process is performed at 50°C for a period of 1 h, the epoxy becomes more suitable. In addition to epoxy, other polymers such as PDMS and SU-8 can also be used in this process. In fact, the key parameters in selecting the polymer as the structure of the microfluidics are as follows:

- The ink filament can preserve its shape during polymer encapsulation. In other words, the mass density of the polymer should be less than the mass density of the ink filament.

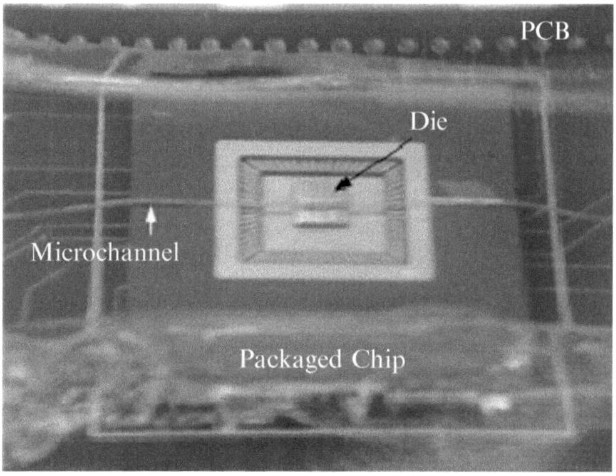

Fig. 5.15 Optical microscope images of a microchannel implemented using the proposed technique in a transparent epoxy on a package and printed circuit board, filled with a coloured liquid for better visualization [306]

- The curing can be performed at a temperature much less than melting point of the ink filament.
- Further specifications of the polymer depend on the dedicated application. For instance, optical transparency is important if the observation of flowing liquid in the channel is required. The thermal conductivity is critical when the heat exchange from the liquid in the channel is necessary. On the other hand, the biocompatibility of the polymer is critical when living bioparticles are injected in microchannel.

Degassing

In addition to the above mentioned issues, the degassing of prepared epoxy is critical for high-precision applications. This is because the presence of an air-bubble a few hundred micrometers in diameter may affect the precision of the microfluidic fabrication process and results in channels with inaccurate dimensions. Figure 5.16a, b shows the role of the degassing process performed under a light vacuum. In these figures, the mixture of epoxy and curing agent prior to and following the degassing process is revealed. Also, the epoxy encapsulation of an ink filament is shown in Fig. 5.16c.

5.3 Direct-Write Microfluidic Packaging Procedure

We elaborate the proposed microfluidic packaging procedure in this section. This procedure consists of six steps which are described below and shown in Fig. 5.17a–f [305].

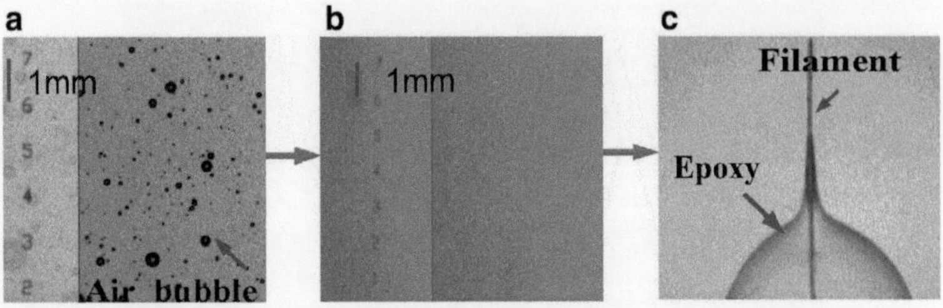

Fig. 5.16 Epoxy preparation and deposition: (**a**) before degassing, (**b**) after degassing, and (**c**) during deposition

5.3.1 Encapsulation of Bonding Pads and Wires

Before starting the three-step direct-write microfluidic fabrication process, the conductors should be covered so as to avoid direct contact with fluids in the channels (Fig. 5.17a). For this, a partially cured epoxy resin (Epon 828, Shell Chemical) is dispensed (Champion 8200 dispenser, Creative Automation Co.) on the packaged chip in order to encapsulate the bonding wires. Due to surface tension and high viscosity of the semi-cured epoxy, it naturally flows around the bonding wires, but stops near the pads (see Fig. 5.18d).

Therefore, the loose die remains uncovered, as is required for sensing purposes. In Fig. 5.18a, the SEM image of a chip (top view) with encapsulated bonding wires is displayed. This chip was covered with a thin layer of gold for better visualization. It should be mentioned that this step can be ignored for a particular arrangement of pads with an open space for ink deposition.

The trajectory points (x, y, z) of ink deposition should initially be measured and programmed into the robot-driven dispensing system. This trajectory path should pass precisely over the sensing electrodes. In order to accomplish this, a high precision optical profiler (WYCO, Veeco, Inc.) should be used to measure the height (z) of the different trajectory points (x, y).

The height profile along the desired trajectory is shown in Fig. 5.18b, c. The envelope of this curve can be used as the z-coordinate of the ink deposition trajectory. It should be mentioned that we have already proposed another manually alignment technique for the same purpose in [140].

5.3.2 Ink Deposition

A paste-like organic ink (mixture of petroleum jelly and a microcrystalline wax [314]) is extruded (Ultra® 2400, EFD Inc.) through a micronozzle and deposited on the substrate. During the extrusion, a micro-positioning robot (Model I&J 2200, I&J FISNAR Inc.) moves the nozzle across a desired trajectory. This sacrificial ink

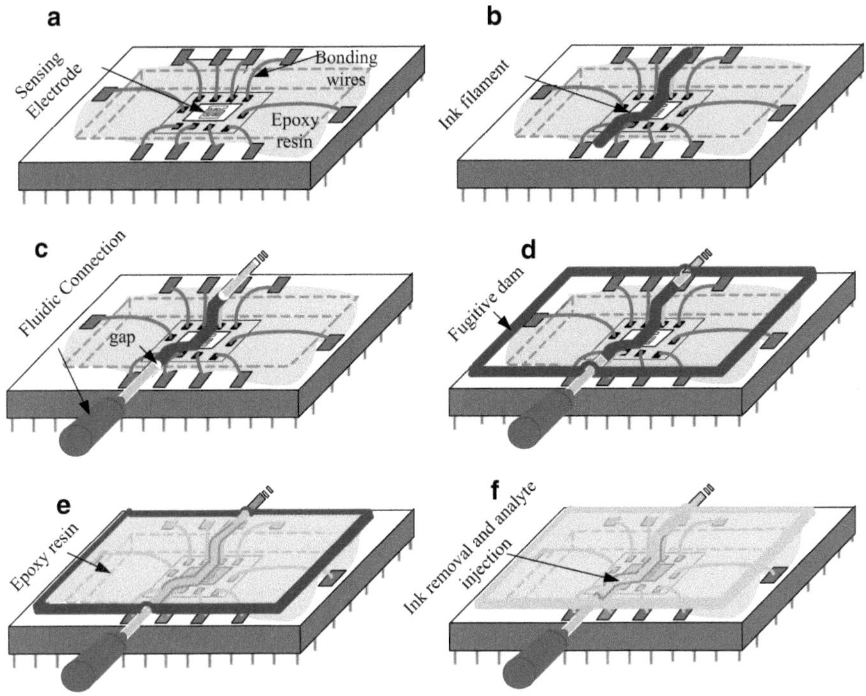

Fig. 5.17 Microfluidic packaging procedure: (**a**) wire bonding encapsulation, (**b**) ink deposition, (**c**) fluidic connection, (**d**) fugitive dam, (**e**) fugitive ink encapsulation, and (**f**) ink removal and analyte injection [306]

structure, shown in Fig. 5.17b, preserves its shape during epoxy encapsulation (Section E). The following parameters should be modulated to control this process: the air-pressure applied to extrude the ink through the micronozzle (P), the velocity of the moving micronozzle over the trajectory (v), the relative height between the nozzle and substrate (H) and the microcrystalline fraction of the organic ink mixture (M). It is worth mentioning that the alignment of the ink filament above the sensing electrode is a very important step in implementing the on-chip microfluidic channel. For many applications, it is not required to use an optical profiler to extract the trajectory of ink deposition. Three-axis robots can be used to find the critical points of the ink trajectory.

Microfluidic fabrication and fluidic connection processes are performed on an electrically packaged CMOS chip. The alignment of the ink deposition process over the desired region of the CMOS chip is performed manually, after the CMOS chip has been fixed to the microrobot platform.

- *Manually alignment*

A small tip needle (10 μm, World Precision Inc.) is moved by the microrobot, set in manual mode, to specific locations over the substrate (e.g. sensing electrodes)

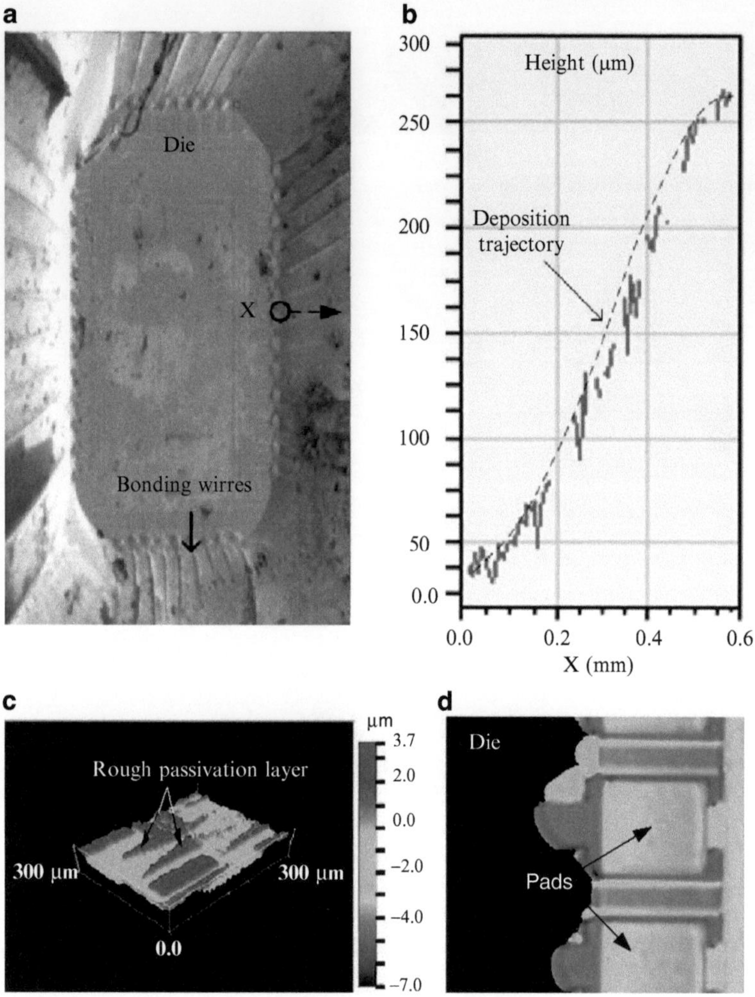

Fig. 5.18 Trajectory recognition for ink deposition: (**a**) top view SEM image of a chip after wire bonding encapsulation, (**b**) height measurements starting from the mark x on the chip, (**c**) laser profiler image of the chip (in the middle of chip), and (**d**) laser profiler image of bonding pads which stop the flowing epoxy resin [306]

while alignment is done visually using a stereomicroscope. The *x*, *y* and *z* coordinates of the points of the deposition trajectory are determined and recorded in the trajectory file. For the fabrication of the straight microchannel shown in Fig. 5.18b, the deposition can be started at one edge of the CMOS chip, crossed over the sensing electrodes and finished at the other edge of the chip. The minimum feature size of this sensing electrode is approximately the same size as the micronozzle tip used during this alignment technique. The observation error E is (D + E)/2 (or ~7 μm, where D is the step size of the robot [140].

- *Alignment error*

The total magnification of the stereomicroscope used (SZX7, Olympus Inc.) is 250X. The minimum observable feature size with the naked eye is assumed to be 1 mm and the observation error *E* is 4 μm (i.e., 1 mm/250). While the micronozzle is moved along one direction to reach a desired location, as shown in Fig. 5.19, two points (i.e. A and B) can be defined at the limit of the position uncertainty of distances $x1$ and $x2$, respectively. The relations between $x1$ and $x2$ are

$$x1 + x2 = D \tag{5.1}$$

$$x1 - x2 < E \tag{5.2}$$

The expressions (5.1) and (5.2) can also be expressed as

$$x1 < (E + D) / 2 \tag{5.3}$$

$$x2 > (D - E) / 2 \tag{5.4}$$

Therefore, the maximum alignment error along one axis is approximately $(D + E)/2$ or 7 μm.

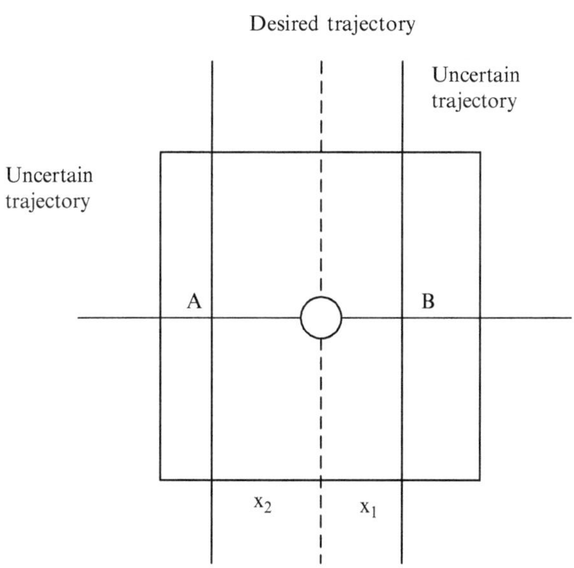

Fig. 5.19 Estimation of alignment error

5.3.3 Fitting Connections

Following the ink deposition process, the microscale fluidic fittings (nozzle or tube) are placed and secured at desired locations close to the deposited ink on the chip using a few drops of hot glue. An extra deposition of fugitive ink from the fitting is necessary to fill the space between the ink filament and fluidic connection and furthermore, to prevent the infiltration of epoxy into the fitting during the encapsulation process (see "gap" in Fig. 5.17c).

5.3.4 Fugitive Dam

Another ink deposition process is performed in the pre-defined boundary of epoxy encapsulation. This fugitive dam can easily be removed during the ink removal step (Fig. 5.17d).

5.3.5 Ink Encapsulation and Filling Process

In this step, a low viscosity epoxy resin is dispensed on the deposited ink within the encapsulation boundary. Curing of the resin occurs at room temperature over 24 h. This epoxy encapsulation process creates a strong and hermetic bond on the uneven surface of the loose die (see Fig. 5.17e). It is obvious that an open-top channel can be performed by using less volume of epoxy. This type of channel is required for post sensing layer deposition, which can be capped afterward.

5.3.6 Ink Removal and Analyte Injection

The fugitive ink is melted at ~75°C and expelled under a light vacuum or air pressure. Hot water is injected through the channel to remove the ink remnants. Just after this step, an analyte solution can be directly injected into the fabricated microchannel, on the microelectronic chip, for sensing purposes (Fig. 5.17f). Following the direct-write microfluidic packaging procedure, a microchannel has been implemented on top of a CMOS sensor, as shown in Fig. 5.20. For this purpose, in order to deposit a fine and continuous 100-μm ink filament, the variable parameters are adjusted to P = 250 kPa, V = 1.5 mm/s, H = 100 μm and M = 40 wt% as reported in [305].

The differential measurements sometime require two microfluidic channels to direct the analyte towards the sensing and reference sites. DWFP can effectively be employed to create such microfuidic channels, as shown in Fig. 5.21a, b. The number of microchannels on the top of CMOS chip can be increased, depending on the

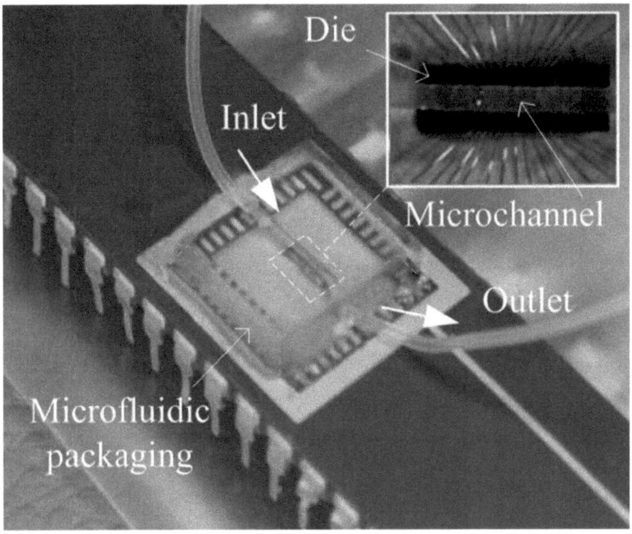

Fig. 5.20 Microphotograph of hybrid test structure including CMOS chip and microfluidic packaging (this platform was embedded in a shielding box during measurement) [66]

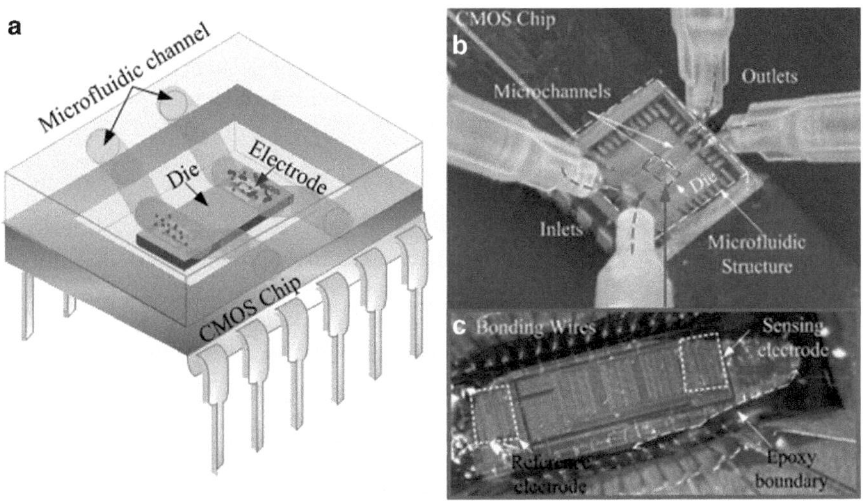

Fig. 5.21 Two microfluidic channels atop CMOS chip: (**a**) illustation of two channels, (**b**) optical microscopic channels, and (**c**) microscopic image of die which is encapsulated by epoxy except sensing and reference electrodes

required channel's inner diameter and the dimensions of the die and electrodes. The first step of the microfluidic packaging procedure (see Fig. 5.18a) has been performed to encapslate everything above the die, except the sensing and reference electrodes , as shown in Fig. 5.21c

5.4 Emerging Applications of DWFP

The Direct-write microfabrication process can be employed to create different microstructures such, as microvalves and optical waveguides, as described in this section.

5.4.1 Microvalve

Microvalves are key devices used to regulate the fluid in a microfluidic system [320]. The principle of conventional microvalves relies on its mechanical or electromagnetic properties which often are neither biocompatible nor suitable for bioassays [321].

Stimuli-response hydrogels, with their excellent biocompatibility, sensing and actuation functions, have become the leading candidate for this purpose. Earlier functional hydrogel-based microvalves were fabricated by in situ photo patterning using photo-sensitive hydrogel. There are different types of hydrogels that respond to external stimuli, such as glucose, antigens, electric field, and temperature [322, 323]. As reported by Wang et al., a cavity should be created to anchor the hydrogel and consequently prevent its movement in the channel [324].

For this, prior to hydrogel formation in microfluidic channels, a machining procedure should be performed to construct such cavities. This chapter presents a new microvalve implementation procedure using DWFP with no need for any further micromachining procedures.

- *Direct-write microvalve*

As already reported by our team, DWFP is an amenable technique to fabricate 3D microfluidic structures (e.g. conic- or cylindrical-shaped chambers [325]) with different aspect ratios. The focus of this section is the implementation of hydrogel-based microvalves by employing such 3D microfluidic components. In this process, similar to moulding techniques, a thermally (or pH) sensitive hydrogel is injected into an implemented microfluidic structure prior to gelation. After sufficient time has

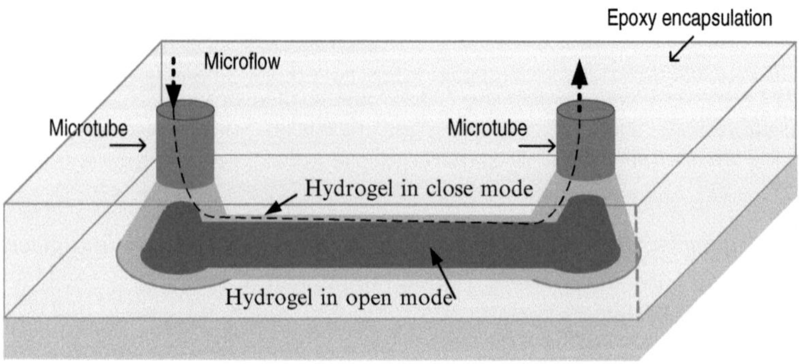

Fig. 5.22 Illustration of proposed direct-write hydrogel-based microvalve

passed, the hydrogel is curred, forming the same shape as the microfluidic structure. Figure 5.22 illustrates the principle of proposed direct-write hydrogel-based microvalve fabrication.

As shown in this figure, a reversible volumetric change property is exploited to open the microvalve while the liquid temperature is greater than the phase transition temperature. It is obvious; the hydrogel will not be pushed out of the microchannel while it is trapped in the two microchambers. Additionally, this figure shows the proposed valve in both open and close mode. In the open mode, the fluids can pass through the channel at low pressures.

Figure 5.23a–d shows five steps to fabricate a direct-write microvalve. These steps are ink deposition, epoxy encapsulation, ink removal, hydrogel (pre-gelation) injection and hydrogel formation. Of course, based on the above discussions, DWFP allows for the implementation of different 3D structures that can be employed for different designs of mocrovalves. The ink architecture and implemented microstructure are shown in Fig. 5.24a, b.

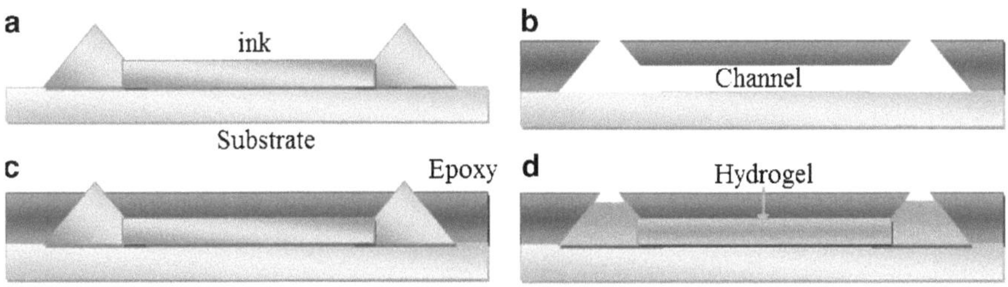

Fig. 5.23 Microvalve fabrication process: (**a**) ink deposition, (**b**) epoxy encapsulation, (**c**) ink removal, and (**d**) pre-gelation hydrogel injection and hydrogel formation

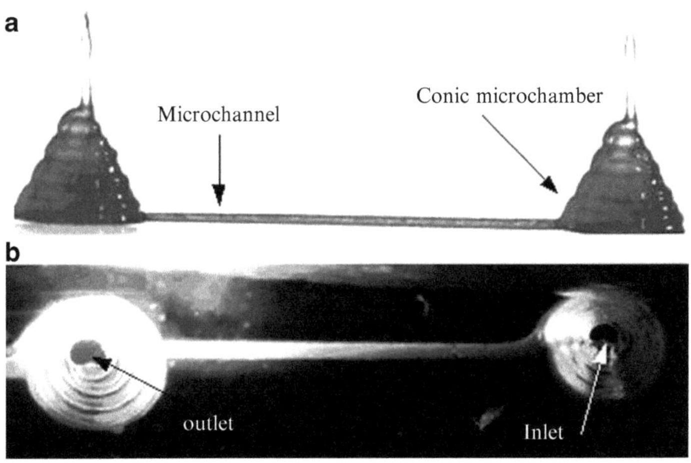

Fig. 5.24 Proposed hydrogel-based microvalve: (**a**) ink architecture and (**b**) corresponding microfluidic structure respectively implemented through DWFP

- *Hydrogel preparation*

The monomer (N-Isopropylacrylamide) is cross-linked with methylene bisacrylamide (N′-methylenebisacrylamide) in the presence of ammonium persulphate as an initiator. The reactants were dissolved in 2 mL of deionised water and the pre-gel solution was mixed with N,N,N′,N′-tetramethylethylenediamine (TEMED) – the accelerator – prior to gelation, and added into the mixture. This mixture was quickly injected into the microchannel. The rate of gelation could be controlled by the concentration of TEMED. After cross-linking overnight (12 h) at room temperature, warm water (T > 32°C) was injected into the microfluidic structure in order to leach out non-reacted reagents in the Poly (NIPAM) gel.

- *Testing results*

The fraction of maximum flow directed through the microvalve is depicted in Fig. 5.25. For this measurement, a weight scale was used to measure the liquid passing through the microchannel for different temperatures of solutions while the fluid indicator (output signal of sensor) was high. As seen in this figure, the microvalve was opened and closed for $T_{max} > 50$ and $T_{min} < 20$, respectively. Of course, with different geometries and chemical recipes, this thermal gap ($T_{max} - T_{min}$) could be minimized for precision LoC applications.

The key parameter of microvalves is the response time. Actually this parameter depends on the geometries of the microvalves, as well as the hydrogel structure. The measured response time is around 34 s using a capacitive sensor that indicates the fluid in the channel.

5.4.2 Direct-Write Heat Exchanger

Thermal management of highly integrated microelectronic devices has become a major issue due to the desire for high power consumption and small packaging size [326].

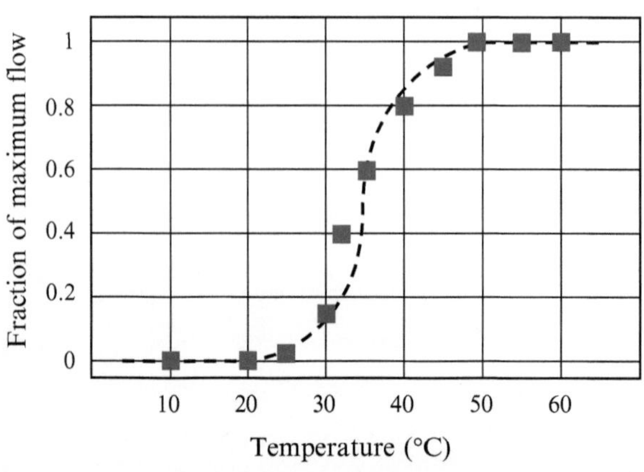

Fig. 5.25 Fraction of maximum flow for different temperatures

As microelectronic chips with embedded micro heat exchangers have good thermal characteristics, the integration of microfluidic channel arrays with microelectonics has recently been highlighted. In addition to LoC applications, DWFP is a good candidate to implement an array of microchannels for cooling microelectronic devices and platforms. The unique advantages of DWFP to construct 3D microfluidic structures on a variety of substrates can efficiently be exploited for cooling purposes [327]. For instance, a fugitive-ink scaffold can be constructed atop PCBs and/or microelectronics (Fig. 5.26). The implemented microfluidic structure atop PCBs or chips with appropriate sealing and further fluidic connections can act as a liquid based heat-sink. Further efforts are required to create capillaries through DWFP by using the nozzles with microscale inner diameters, rather than the commercial 100-μm nozzles currently available.

5.4.3 Optical Waveguide for Biosensing Applications

Arrays of optically transparent ridge-type waveguide microstructures, used for optical bio-imaging applications, can be fabricated using various techniques, including conventional micromachining procedures, as well as DWFP [306, 328, 329]. This section presents the development of polymeric waveguides with a strong adhesion to a wide range of substrates (e.g. silicon, plastic and glass) with features of the order of 5 μm in width and up to several millimetres in length. The implementation of such waveguides is a unique advantage of direct-write assembly techniques for large-area analysis in cell cultures and tissue imaging. As already mentioned, another advantage of this technique, as compared to conventional photolithographic or soft-microfabrication techniques, is the dynamic and easy variability of multi-dimensional structures and their fabrication, which does not require a mask to pattern, define and etch the desired structures.

The waveguide structures can be used as a support platform for immobilizing biochemical recognition elements, such as sol–gel-derived xerogel-based thin films. In order to pattern sol–gel sensing materials, DWFP offers a simple yet powerful method to create complex three-dimensional microstructures having the combined versatile capabilities of biological and chemical sensing, fluidic sample distribution and transport for sensor arrays, and integration with CMOS integrated

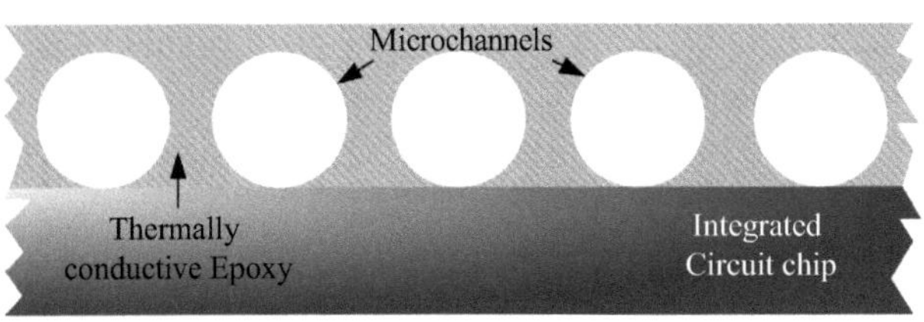

Fig. 5.26 Illustration of a direct-write heat exchanger

circuits for signal detection and processing. As shown in Fig. 5.27, a light emitting diode is employed to excite functionalized sol–gel layer impregnated with oxygen sensitive luminophore *tris*(2,2′-bipyridyl) dichlororuthenium(II)hexahydrate ($[Ru(bpy)_3]^{2+}\cdot 6H_2O$) [306]. The sol–gel sensing layer is immobilized on the optical polymer waveguides. As a consequence, the amount of fluorescence generated corresponds to the O_2 concentration, and is captured by an optical filter and finally guided towards the photodetector. The interface circuit detects and transmits the signal accordingly.

- *Fabrication process*

The fabrication process is very similar to the microfluidic fabrication processes discussed in previous sections that are to create microchannels. In fact, for microfluidic purposes, the extruded organic ink filament is encapsulated with epoxy resin and thereafter removed in order to create a hollow structure. This organic ink is deposited in a desired trajectory to construct a casting which is filled with epoxy resin. This ink can thermally be removed after epoxy curing, leaving behind micropatterned epoxy or polymeric structures, as shown in Fig. 5.28a–d. Xerogel waveguides are also dispensed in between the polymeric waveguides. A class II hybrid xerogel can be incorporated using a direct-write technique for O_2 monitoring purposes. In order to accomplish this, the epoxy surface was first coated with the adhesion promoter, 3-aminopropyltriethoxysilane (APTS). The APTS was then dried at 50°C for 2 h. Once dry, the epoxy surface was again cleaned with isopropanol. The prepared sol–gel sensing layer could subsequently be deposited by direct-dispense in the spaces in between the two polymeric waveguide structures. It should be mentioned that the xerogel recognition element was prepared by encapsulating the luminophore, (tris(4,7-diphenyl-1,10-phenathroline)ruthenium(II)) ([Ru(dpp)3]2+) in the xerogel nanoporous matrices [306]. The sensor waveguide structures should be dried at 50°C for 5 days.

- *Experimental results*

The optical microscopic images of fugitive casting as well as polymeric waveguide are shown in Fig. 5.29a, b. The optical waveguides with rigid structure were fabricated using a DWFP (Fig. 5.29b). Figure 5.30 shows the preliminary measurement

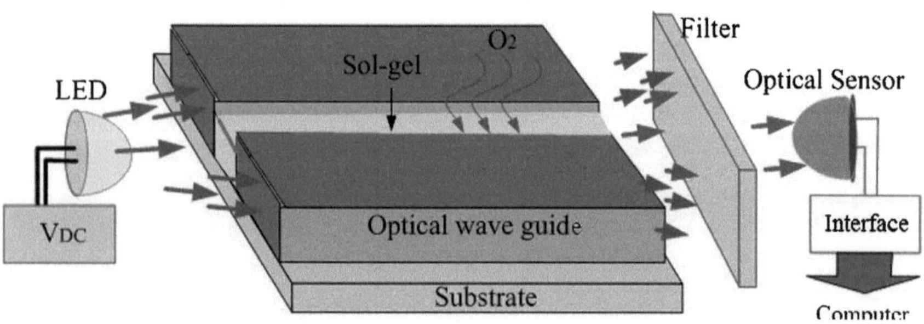

Fig. 5.27 Simplified illustration of proposed optical measurement system with optical waveguide

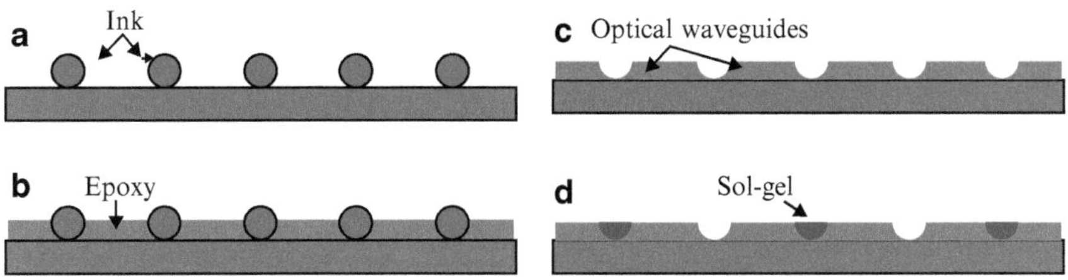

Fig. 5.28 Fabrication process of proposed optical measurement system through DWFP and sole gel immobilization process: (**a**) ink deposition, (**b**) epoxy infiltration, (**c**) ink removal, and (**d**) sol–gel deposition

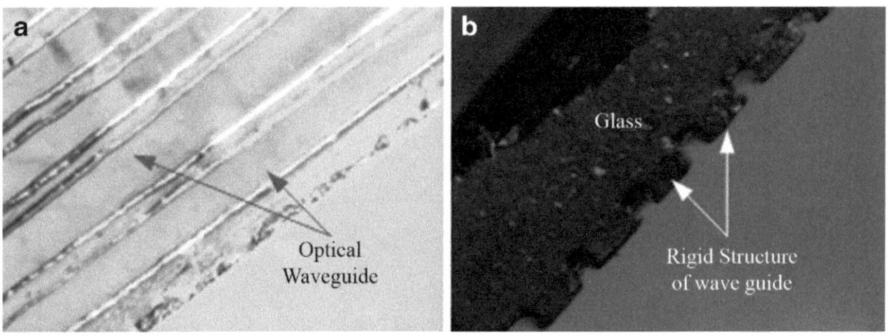

Fig. 5.29 Optical microscopic images of implemented optical waveguide with immobilized sol–gel: (**a**) from top of chip, and (**b**) cross section (D_Z = 250 μm)

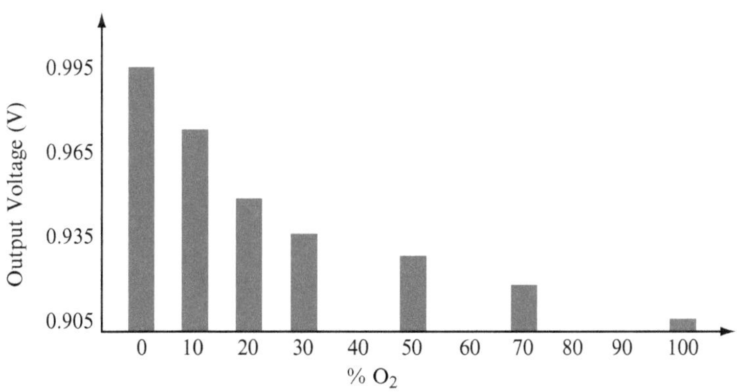

Fig. 5.30 Experimental results of optical sensor using DWFP

results where the optical sensor chip was exposed to O_2 with different concentrations. A similar sensing platform can be employed in the near future for other applications, such as glucose detection, using the appropriate functionalized sol–gel layer.

5.5 Summary

We described the recent published microfluidics packaging methods for CMOS based Lab-on-Chip applications. Among these, direct-write microfluidic packaging technique was introduced as CMOS-compatible hermetic method for biosensors. Based on the discussed issues throughout this chapter, DWFP can effectively be employed to fabricate microfluidic devices such as channel, valve, chamber or mixers. Further modifications should be applied over this technique to implement high precision microfluidic structures using micro/nanoscale nozzle.

Chapter 6
Current Technology and Future Works

The capacitive measurements of deposable sensing electrodes are conventionally performed using Electrochemical Impedance Spectroscopy (EIS) [331, 332]. As shown in Fig. 6.1a, such a measurement device is connected to an array of electrodes which are exposed to analyte. Recently handheld EIS systems have received much attention as opposed to conventional EIS systems. A handheld system featuring an array of sensing sites (Fig. 6.1b) can be used for several point-of-care applications such as blood analysis (e.g. minilab, Abaxis Inc. [333]) or environmental monitoring such as bacteria detection [334]. However, researchers involved in circuit and system design and relevant biotechnological studies are willing to embed such portable systems in a single chip in the near future. In this direction, a CMOS based capacitive sensing LoC can be implemented in a syringe style package as shown in Fig. 6.1c. The biological or chemical analyte is directed by syringe towards the sensing sites through the nozzle. After each measurement, the nozzle and sensing site will be cleaned using the appropriate solutions which are directed into the channel and sensing sites in the same manner.

6.1 Conventional Impedometric and Capacitive Measurement Systems

EIS is a well established technique routinely used in biological or chemical laboratories and is based on measuring the behaviour in which a sinusoidal electric current or voltage signal at different frequencies is conducted through a sample under test. EIS is used for a variety of applications such as for detecting DNA, viruses, glucose and lactate [331, 332].

There are many table-top impedance measurement systems commercialized by several companies as can be seen in Table 6.1.

This table introduces the commercialized impedance measurement devices with Frequency Response Analysis (FRA) or Continuous Time (CT) manners.

E. Ghafar-Zadeh and M. Sawan, *CMOS Capacitive Sensors for Lab-on-Chip Applications: A Multidisciplinary Approach,* Analog Circuits and Signal Processing,
DOI 10.1007/978-90-481-3727-5_6,

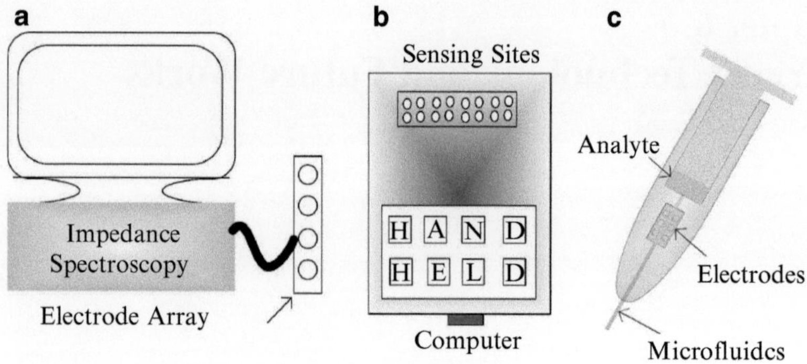

Fig. 6.1 Capacitive biosensing system (**a**) conventional, (**b**) current, and (**c**) future technologies

Table 6.1 Commercialized impedance measurement systems

Model	Company	Mode
Alpha-A	Novocontrol	FRA
Reference 600	Gamry Instruments	FRA
LEIS370	Princeton Applied Research	FRA
1255A	Solartron Analytical	FRA
Model 3120	Venable Instruments	FRA
Model 2505	Clarke-Hess Communications Research	FRA
SA Series 01	Core Technology Group	FRA
ECIS Z	Applied Biophysics	CT
96X	ACECS Bioscience	CT

The measurement systems provided by Applied Biophysics and ACEA Biosciences Incorporations can be used for continuous time multi-sample analysis.

In addition to the aforementioned EIS systems, non-conventional measurement systems have also be used in biological laboratories. Figure 6.2 shows an impedometric measurement system set-up reported for bacteria growth monitoring [71]. In this measurement set-up, a lock-in amplifier is used to synthesize a sinusoidal signal on the sensing electrodes. A user interface software (UIS) is used to select the frequency, the voltage amplitude and the measurement time period. A multiplexer is also prepared to select the electrode through the UIS. An array of electrodes provided from Applied Physics Inc. is also seen in this figure.

The electrical modeling of the electrodes exposed to analyte (e.g. bacterial solution) can be represented as a serial combination of the sample medium resistance (R_s) and two double layer capacitances (C_{dl1}, C_{dl2}) in proximity of the electrodes E_1 and E_2 as shown in Fig. 6.3a, b [71]. By assuming $C_{dl1} = C_{dl2} = C_{dl}$, the impedance of this sensing electrode can roughly be expressed as

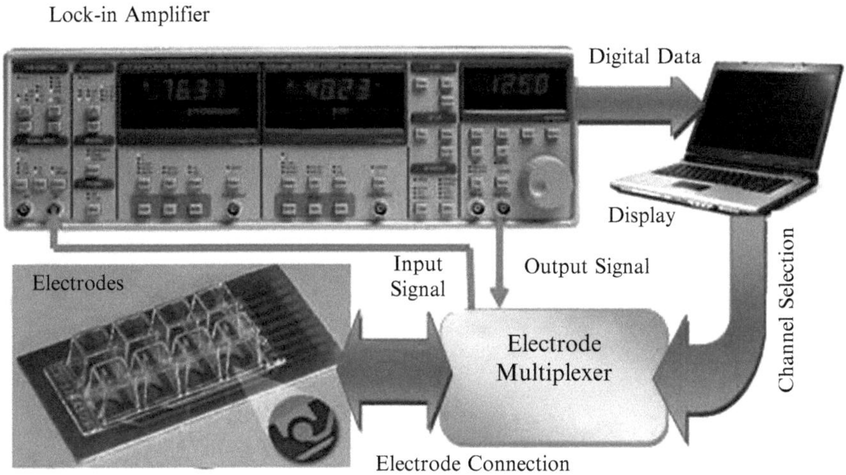

Fig. 6.2 A non-conventional impedance measurement system for cell growth monitoring

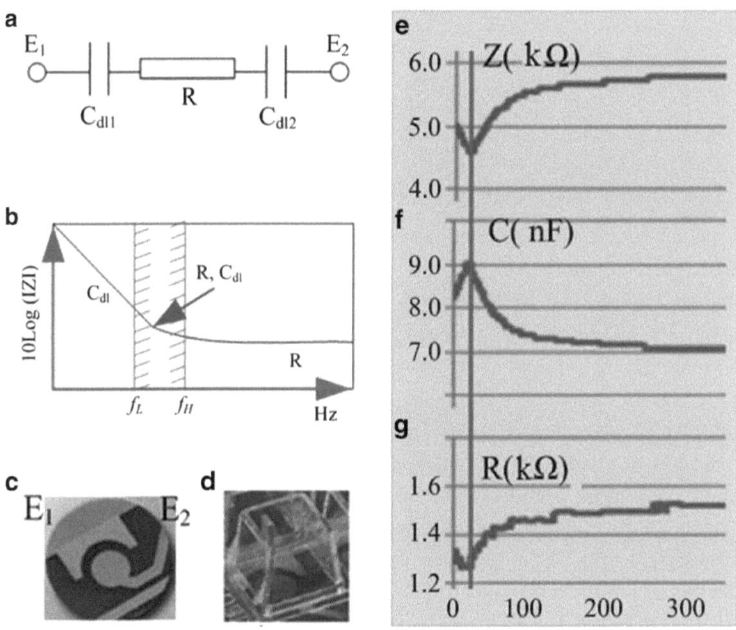

Fig. 6.3 Impedance-based bacteria growth: (**a**) electrical model of electrodes, (**b**) diagram of a frequency response, (**c**) electrode integrated with a well provided by Applied Physics Inc., (**d**) representation of well, (**e**) bacteria growth monitoring through impedance, (**f**) capacitance and (**g**) resistance measurements

$$Z = R_S - \frac{j}{f C_{dl}} \qquad \sqrt{-1} \tag{6.1}$$

where j = $\sqrt{-1}$ and f is the frequency of the applied sinusoidal signal. As illustrated in Fig. 6.3c, the capacitance C_{dl} is the dominant component at lower frequencies where the sample medium resistance R_s could be effectively ignored. At high frequencies (see Fig. 6.3c), however, the dominant contribution to the total impedance is R_s.

Of course, the cut-off frequency $f_L = 1/(2\pi R_S C_{dl})$ is a function of capacitance and the value of this capacitance depends on many factors including applied sinusoidal voltage, and electrode properties (e.g. electrode material, roughness, geometry) [335, 336]. As the value of applied voltage and temperature are fixed during the laboratory experiments, their effects can be ignored and the detected signal can only be considered as a function of bacteria growth. The monitoring of biological activities (e.g. bacteria growth [337, 338]) can also be monitored by R_s or preferably C_{dl} and not necessarily from a combination of both of them as shown in Fig. 6.3e–g.

Figure 6.3e–g shows that the growth of *E coli* bacteria with a concentration of 10^7 in 1 mL LB causes a change in the capacitance and resistance of electrode (ECIS, Applied Physics Inc.) incubated at 37.5°C. Figure 6.2d also shows a centimeter-scale well for sample handling purposes while as mentioned in Chapter V, a direct-write microfluidic packaging technique can effectively be employed to create the microchannels [304]. Based on the above discussion, the capacitance variation can be extracted from the impedance measurement results.

6.2 Handheld Impedance Measurement Systems

A handheld impedance or capacitance measurement system consists of a disposable electrode that is fixed on the handheld system (e.g. Carbon printed electrode) and an impedance reader card such as the AD5933 (Fig. 6.4a) and USB-6281 (Fig. 6.4b) which are made by Analog Devices and National Instruments respectively. For example, AD5933 has recently been reported for many applications including health monitoring, single cell analysis and glucose monitoring [339–341].

The impedance reader card can be integrated with other microelectronic circuitry to readout from a large array of electrodes. For instance AD5933 was used to measure the impedances associated with a number of electrodes for protein-based biosensing purposes [325]. Figure 6.5 shows the circuitry dedicated for connecting N number of electrodes to a single-input impedance reader.

As shown in Fig. 6.5, this measurement system consists of three different components: an impedance coder (IMC), an impedance reader chip (IRC) and an impedance decoder (IMD). IRC is employed to convert the impedance variation of a single electrode to digital values. Special codes are added to the impedance signal

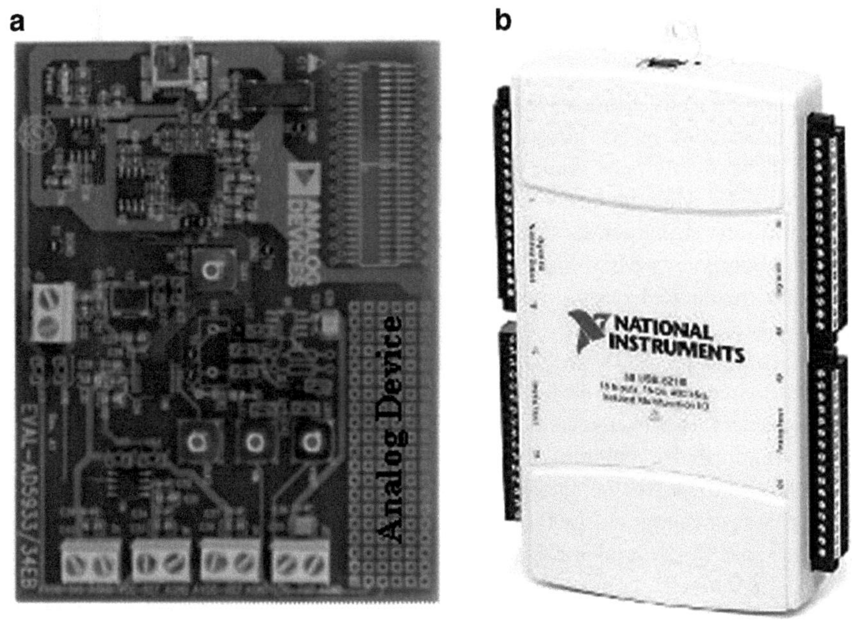

Fig. 6.4 Handheld impedance measurement systems provided by (**a**) Analog device Inc. and (**b**) National Instrument Inc.

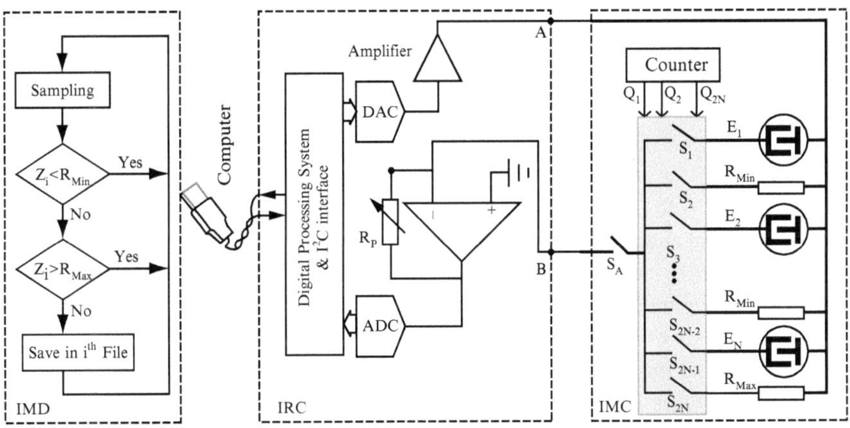

Fig. 6.5 Handheld impedance/capacitive biosensing system array

in order to identify the location of the sensing electrode in the array. This impedance signal is thereafter converted to digital values which are transferred into computer through IRC for decoding and signal processing purposes. The recorded data is broken into several parts by an IMD algorithm where each part relates the impedance values to a specific electrode.

6.3 Towards Fully Integrated Capacitive Sensing LoC

An on-chip capacitive biosensing LoC includes an application specific integrated circuit (ASIC), a microfluidic structure and biofunctionalized sensing electrodes. Several microelectronic integrated circuit manufacturers such as Taiwan Semiconductor Manufacturing Company (TSMC) produce the CMOS chips and the micofluidic structures are commercialized by many microfluidic manufacturers such as Micralyne or Micronit companies [342, 343]. However, still neither microfluidic nor the microelectronic manufacturers are offering hybrid CMOS microfluidic chips for LoC applications.

The standard CMOS technology serves as the chassis on which various integrated circuitries are built for a variety of applications such as image sensors and accelerometers. However several more steps should be taken to create a hybrid technology platform for emerging Lab-on-Chip applications in the future. Current technology allows for the integration of CMOS chip and microfluidic structure through non-conventional methods including packaging methods as described in chapter V [306]. Several modifications should be applied to current packaging methods in order to control the microfluidic packaging process. Further efforts should also be directed toward CMOS capacitive biosensing designs in order to create the reliable, highly sensitive, repeatable and miniaturized devices for future use. Herein some important issues are described.

6.3.1 Packaging

A microelectronic device only includes some input and output electrical signals, while in a CMOS LoC, in addition to electrical signals, there are some fluidic inlets and outlet (Fig. 6.6a). For instance, power supply and analyte are applied to a CMOS chip as electrical and fluidic inputs respectively where the ground and the connection to waste reservoir are the electrical signal and fluidic outputs respectively. The electrical packaging of CMOS chip is performed using standard techniques such as flip-chip and wire bonding where the microfluidic packaging are performed through non-conventional methods such as direct-write method. Of course, standard microfluidic packaging is the main requirement for CMOS based LoCs. This packaging method should be CMOS compatible, repeatable and reliable with hermetic bonding. In Chapter V, we discussed the advantages of direct-write for these purposes. However, further studies are required in order to standardize this technique.

6.3.2 Capacitance Characterization

The provided CAD tools for the capacitance characterization are applicable for the microelectronic devices in deep CMOS chip. For example the coupling capacitance between two conductors can be calculated precisely, but the parasitic

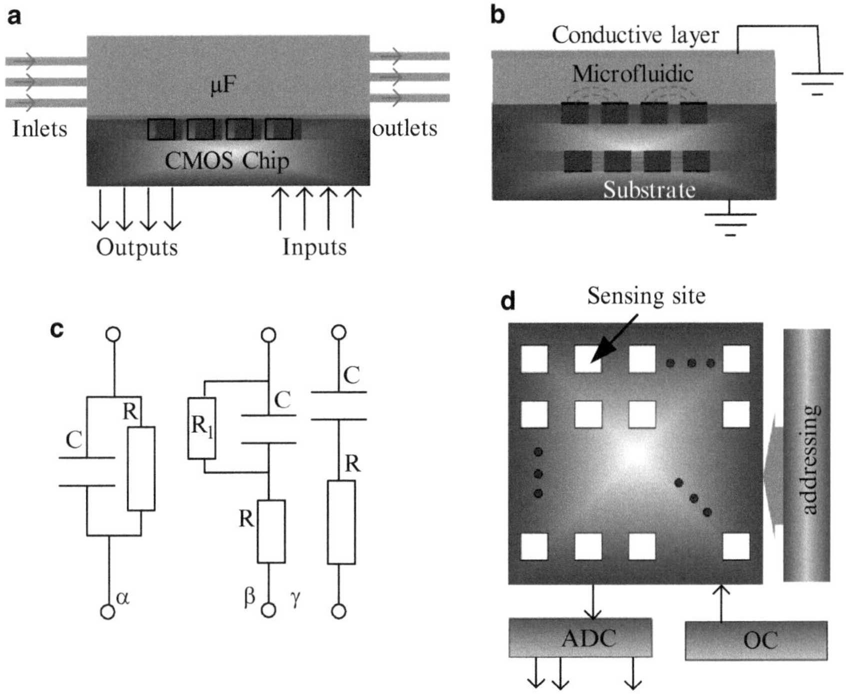

Fig. 6.6 Illustrations of future works on CMOS based LoC: (**a**) packaging, (**b**) capacitive characterization, (**c**) electrical models (α, β, γ), and (**d**) generic capacitive sensing chip (OC: Offset cancellation)

capacitance between the topmost metal layer and the substrate can not be measured particularly for cases while complicated configurations (see Chapter 2) are used. The minimum requirement in the design of a capacitive sensor is the estimation of the parasitic capacitance created by the sensing electrode above the CMOS chip. Furthermore, the principle of standard capacitance characterization methods is based on the created capacitances between each two conductors and between each conductor and the grounded substrate. However, in many of LoC applications, the microfluidic structure is covered by a grounded metal layer made from indium tin oxide (ITO) or other appropriate materials. Therefore, the characterization of a CMOS-based capacitive sensing LoC is another important unmet challenging issue.

6.3.3 Electrical Modeling of Biological Sample

The CAD tools used for the design of microelectronic circuitries are implemented based on very accurate models provided for CMOS transistors and other microelectronic devices. Of course, the design of capacitive biosensor on CMOS chips could not be possible without exact models of biological samples and/or recognition element formed on the top of the chip. Electrochemical models of the aforementioned capacitive biointerfaces have been reported for various technical circumstances.

An electrical model of biological and/or chemical samples is a circuit network of some resistors and capacitors (see Fig. 6.6c). This circuit topology and the dynamic range of each resistor or capacitor is influenced by many factors including the material of the topmost metal or passivation layers, the geometry and dimensions of electrodes, the concentration of the target molecules or cells and non-specific molecules or cells which might affect the dielectric property of analyte in proximity of the sensing electrodes. CMOS-based electrochemical models of capacitive biointerface are another important issue which should be studied in the future. Through these models, the design of capacitive sensor can be performed by electronic designers using new modified CAD tools.

6.3.3.1 Generic Microelectronic Circuitry

Previously, several papers have been published on the design of generic capacitive sensor system for MEMS based applications such as accelerometers, but as already mentioned in Chapter 4, the capacitive sensing LoCs should be designed and implemented using a different design strategy. The design of a generic capacitive sensor LoC is another important challenge for microelectronic designers. An optimised generic system should feature a large array of capacitive sensors, offset cancellation and calibration module and, a high resolution low speed ADC preferably using a sigma delta modulation technique (Fig. 6.6d).

6.3.4 Cleaning Procedure

In a standard biological laboratory, only disposable electrodes are used to prevent transmission of infectious agents. These electrodes have already been sterilized and are ready-to-use with no necessity of further cleaning. However, an integrated capacitive sensor should be employed for several measurements. In general, if non-disposable devices are employed, they should be suitably cleaned before and after each use. A cleaning procedure should be designed based on the biochemical interaction of analyte with the sensor. A cleaning procedure applied for biological experiments might not be suited for other experiments.

6.4 Summary

Past, present and future technologies for the capacitance measurement of biological and chemical samples were briefly described in this chapter. Based on the issues discussed throughout this book and in particular this chapter, it can be seen that many techniques are being developed to for CMOS capacitive biosensors, however CMOS based capacitive sensing LoC is in its early stage of developments and further studies are required to design and implement generic capacitive sensor for a wide range of biological applications.

References

1. C.D. Chin, V. Linder, S.K. Sia, Lab-on-a-chip devices for global health: Past studies and future opportunities. Lab Chip **5**(1), 41–57 (2007)
2. H. Craighead, Future lab-on-a-chip technologies for interrogating individual molecules. Nature **442**(7101), 387–393 (2006)
3. A. Chow, Lab-on-a-chip: Opportunities for chemical engineering. AIChE J **48**(8), 1590–1595 (2002)
4. R.S. Tuan, C.W. Lo, *Developmental Biology Protocols: Methods in Molecular Biology*, vol. I (Human Press Inc., Totowa, New Jersey, 2000)
5. R.S. Tuan, C.W. Lo, *Developmental Biology Protocols: Methods in Molecular Biology*, vol. II (Human Press Inc., Totowa, New Jersey, 2000)
6. R.S. Tuan, C.W. Lo, *Developmental Biology Protocols: Methods in Molecular Biology*, vol. III (Human Press Inc., Totowa, New Jersey, 2000)
7. J.R. Harris, J. Graham, D. Rickwood, *Cell Biology Protocols* (Wiely, West Sussex, England, 2006)
8. R.P. Hertzberg, A.J. Pope, High-throughput screening: new technology for the 21st century. Curr. Opin. Chem. Biol. **4**(4), 445–451 (2000)
9. P. Abgrall, A.-M. Gu, Lab-on-chip technologies: making amicrofluidic network and coupling it into a complete microsystem-a review. J. Micromech. Microeng. **17**(5), 15–49 (2007)
10. D.J. Laser, J.G. Santiago, A review of micropumps. J. Micromech. Microeng. **14**(6), 35–64 (2004)
11. K.W Oh, C.H Ahn, A review of microvalves. J. Micromech. Microeng. **16**(5), 13–39 (2006)
12. A.J. Tüds, G.A.J. Besselink, R.B.M. Schasfoort, Trends in miniaturized total analysis systems for point-of-care testing in clinical chemistry. Lab Chip **1**(2), 83–95 (2001)
13. B.L. Ziober, M.G. Mauk, E.M. Falls, Z. Chen, A.F. Ziober, H.H. Bau, Lab-on-a-chip for oral cancer screening and diagnosis. Head Neck **30**(1), 111–121 (2008)
14. W.P. Guo, X.M. Ma, Y. Zeng, Clinical laboratories on a chip for human immunodeficiency virus assay. 27th Annual International Conference of IEEE-EMBS, Shanghai, China, 17–18 Jan 2005
15. D. Yang, L. Song, Y. Liu, R. Zhao, L. Ma, Y. Shao, X. Jiang, A fast, high throughput, and low-cost microfluidic bioassays for detecting HIV. *Proceedings of the 5th International Conference on Information Technology and Application in Biomedicine*, Shenzhen, China, 30–31 May 2008
16. C.P. Price, Point of care testing. Br. Med. J. **322**, 1285–1288 (2001)
17. P.V. Lode, Point-of-care immunotesting: Approaching the analytical performance of central laboratory methods. Clin. Biochem. **38**(7), 591–606 (2005)
18. C.S. Bradshaw, A.N. Morton, S.M. Garland, L.B. Horvath, I. Kuzevska, C.K. Fairley, Evaluation of a point-of-care test, BVBlue, and clinical and laboratory criteria for diagnosis of bacterial vaginosis. J. Clin. Microbiol. **43**(3), 1304–1308 (2005)

19. K. Dickstein, A. Cohen-Solal, G. Filippatos et al., ESC guidelines for the diagnosis and treatment of acute and chronic heart failure. Eur. J. Heart Fail. **10**, 933–989 (2008)
20. P.K.S. Lam, Use of biomarkers in environmental monitoring. Ocean Coastal Manage. **52**(7), 348–354 (2009)
21. P. Hansen, *Advanced Environmental Monitoring, Chapter 23, Biosensors for Environmental Monitoring* (Springer, Netherland, 2007)
22. M. Kampa, E. Castanas, Human health effects of air pollution. Environ. Pollut. **151**(2), 362–367 (2008)
23. J. Melin, S.R. Quake, Microfluidic large-scale integration: the evolution of design rules for biological automation. Ann. Rev. Biophy. Biomol. Struct. **36**, 213–231 (2007)
24. D. Sparks, R. Smith, M. Straayer, J. Cripe, R. Schneider, A. Chimbayo, S. Anasari, N. Najafi, Measurement of density and chemical concentration using a microfluidic chip. Lab Chip **3**, 19–21 (2003)
25. Section of Cancer Genomics. Protocol: DNA preparation from Blood http://www.riedlab.nci.nih.gov/publications/DNA%20Prep_Blood%20.pdf.
26. C.H. Mastrangelo, M.A. Burns, D.T. Burke, Microfabricated devices for genetic diagnostics. Proc. IEEE **86**(8), 1769–1787 (1998)
27. J.W. Hong, S.R. Quake, Integrated nanoliter systems. Nat. Biotech. **21**, 1179–1183 (2003)
28. P. Yager, T. Edwards, E. Fu, K. Helton, K. Nelson, M.R. Tam, B.H. Weigl, Microfluidic diagnostic technologies for global public health. Nature **442**, 412–418 (2006)
29. P. Mitchel, Microfluidics – downsizing large-scale biology. Nature **19**, 217–221 (2001)
30. A.L. Paguirigan, D.J. Beebe, Microfluidics meet cell biology: bridging the gap by alidation and application of microscale techniques for cell biological assays. Bioassays **30**, 811–821 (2008)
31. P. Maoa, J. Han, Fabrication and characterization of 20 nm planar nanofluidic channels by glass–glass and glass–silicon bonding. Lab Chip **5**, 837–844 (2005)
32. B.R. Flachsbart, K. Wong, J.M. Iannacone, E.N. Abante, R.L. Vlach, P.A. Rauchfuss, P.W. Bohn, J.V. Sweedlerbc, M.A. Shannon, Design and fabrication of a multilayered polymer microfluidic chip with nanofluidic interconnects via adhesive contact printing. Lab Chip **6**, 667–674 (2006)
33. S.G. Im, K.W. Bong, C.H. Lee, P.S. Doyle, K.K. Gleason, A conformal nano-adhesive via initiated chemical vapour deposition for microfluidic devices. Lab Chip **9**, 411–416 (2009)
34. H. Becker, U. Heim, Hot embossing as a method for the fabrication of polymer high aspect ratio structures. Sens. Actuat. A: Phys. **83**(1), 130–135 (2000)
35. G. Kumar, H.X. Tang, J. Schroers, Nanomoulding with amorphous metals. Nature **457**, 868–872 (2009)
36. Micronit Microfluidics Inc., http://www.micronit.com/
37. D. Sung Yoon, Y. Lee, Y. Lee, H.J. Cho, S.W. Sung, K. WOh, J. Cha, G. Lim, Precise temperature control and rapid thermal cycling in a micromachined DNA polymerase chain reaction chip. J. Micromech. Microeng. **12**, 813–823 (2002)
38. J.A. Higgins, S. Nasarabadi, J.S. Karns, D.R. Shelton, M. Cooper, A. Gbakima, R.P. Koopman, A handheld real time thermal cycler for bacterial pathogen detection. Biosens. Bioelectron. **18**, 1115–1123 (2003)
39. Harvard Apparatus. http://www.harvardapparatus.com
40. J.B. Christen, A.G. Andreou, Design, fabrication, and testing of a hybrid CMOS/PDMS microsystem for cell culture and incubation. IEEE Trans. Biomed. Circuits Syst. **1**(1), 3–18 (2007)
41. W.E. Wright, J.W. Shay, Inexpensive low-oxygen incubators. Nat. Protoc. **1**, 2088–2090 (2006)
42. A.J. deMello, DNA amplification moves on. Nature **422**, 19–20 (2003)
43. eppendrof Inc. http://www.eppendorfna.com
44. Institute of Bioengineering and Nanotechnology. http://www.ibn.a-star.edu.sg/
45. T.B. Jones, Basic theory of dielectrophoresis and electrorotation. IEEE Mag. Eng. Med. Biol. **22**(6), 33–42 (2003)
46. B.H. Lapizco-Encinas, M. Rito-Palomares, Dielectrophoresis for the manipulation of nanobioparticles. Electrophoresis **28**, 4521–4538 (2007)

47. J. Korlach, C. Reichle, T. Muller, T. Schnelle, W.W. Webb, Trapping, deformation, and rotation of giant unilamellar vesicles in octode dielectrophoretic field cages. Biophys. J. **89**, 554–562 (2005)
48. N. Flores-Rodringuez, G.h. Markx, Improved levitation and trapping of particles by negative dielectrophoresis by the addition of amphoteric molecules. J. Phys. D: Appl. Phys. **37**, 353–361 (2004)
49. Y. Kang, D. Li, S.A. Kalams, J.E. Eid, DC-dielectrophoretic separation of biological cells by size. Biomed. Microdevices **10**, 243–249 (2008)
50. Youlan Li, C. Dalton, H. John Crabtree, G. Nilsson, K.V.I.S. Kaler, Continuous dielectrophoretic cell separation microfluidic device. Lab Chip **7**, 239–248 (2007)
51. P.R.C. Gascoyne, J.V. Vykoukal, Dielectrophoresis-based sample handling in general-purpose programmable diagnostic instruments. Proc. IEEE **92**(1), 22–42 (2004)
52. V. Dolnik, DNA sequencing by capillary electrophoresis. J. Biochem. Biophy. Meth. **41**, 103–119 (1999)
53. T. Rabilloud, Two-dimensional gel electrophoresis in proteomics: old, fashioned, but it still climbs up the mountains. Proteomics **2**, 3–10 (2002)
54. Caliper Inc. http://www.caliperls.com/
55. D. Proudnikov, A. Mirzabekov, Chemical methods of DNA and RNA fluorescent labelling. Nucleic Acids Res. **24**(22), 4535–4542 (1996)
56. G. Kurlyandskaya, V. Levit, Magnetic Dynabeads® detection by sensitive element based on giant magnetoimpedance. Biosens. Bioelectron. **20**, 1611–1616 (2005)
57. M. Khine, C. Ionescu-Zanetti, A. Blatz, L.P. Wang, L.P. Lee, Single-cell electroporation arrays with real-time monitoring and feedback control. Lab Chip **7**, 457–462 (2007)
58. D. Ham, H. Lee, R. Westervelt, *CMOS Biotechnology* (Springer, Berlin, 2008)
59. L. Pancheri, M. Scandiuzzo, D. Stoppa, G.F.D. Betta, Low-noise avalanche photodiode in standard 0.35-μm CMOS technology. IEEE Trans. Electron Devices **55**(1), 457–461 (2008)
60. C. G. Jakobson, U. Dinnar, M. Feinsod, Y. Nemirovsky, Ion-sensitive field-effect transistors in standard CMOS fabricated by post processing. IEEE J. Sens. **2**(4), 279–287 (2002)
61. M.A.P. Pertijs, G.C.M. Meijer, J.H. Huijsing, Precision temperature measurement using CMOS substrate PNP transistors. IEEE J. Sens. **4**(3), 294–300 (2004)
62. G. Medoro, C. Nastruzzi, R. Guerrieri, R. Gambari, N. Manaresi, Lab on a chip for live-cell manipulation. Des. Test Comp. IEEE **24**(1), 26–36 (2007)
63. Hakho Lee, Yong Liu, R.M. Westervelt, D. Ham, IC/microfluidic hybrid system for magnetic manipulation of biological cells. IEEE J. Solid State Circuits **41**(6), 1471–1480 (2006)
64. A. Hassibi, T.H. Lee, A Programmable 0.18-$muhbox{m}$ CMOS electrochemical sensor microarray for biomolecular detection. IEEE J. Sens. **6**(6), 1380–1388 (2006)
65. M.M. Ahmadi, G.A. Jullien, A wireless-implantable microsystem for continuous blood glucose. IEEE Trans. Monit. Biomed. Circuits Syst. **3**(3), 169–180 (2009)
66. E. Ghafar-Zadeh, M. Sawan, A hybrid microfluidic/CMOS capacitive sensor dedicated to lab-on-chip applications. IEEE Trans. Biomed. Circuits Syst. **1**(4) (2007)
67. P.A. Hammond, D. Ali, D.R.S. Cumming, Design of a single-chip pH sensor using a conventional 0.6-μm CMOS process. IEEE Sens. J. **4**(6), 706–712 (2004)
68. L. Yue, C. Vancura, K.U. Kirstein, J. Lichtenberg, A. Hierlemann, Monolithic resonant-cantilever-based CMOS microsystem for biochemical sensing. IEEE Trans. Circuits Syst. I **55**(9), 2551–2560 (2008)
69. H. Eltoukhy, K. Salama, A.E. Gamal, A 0.18-μm CMOS bioluminescence detection lab-on-chip. IEEE J. Solid State Circuits **41**(3), 651–662 (2006)
70. A. Balasubramanian, B. Bhuva, R. Mernaugh, F.R. Haselton, Si-based sensor for virus detection. IEEE J. Sens. **5**(3), 340–344 (2005)
71. E. Ghafar-Zadeh, M. Sawan, V.P. Chodavarapu, Micro-organism-on-chip: emerging direct-write CMOS-based platform for biological applications. IEEE Trans. Biomed. Circuits Syst. **3**, 212–219 (2009)
72. C. Stagni, C. Guiducci, L. Benini, B. Ricco, S. Carrara, C. Paulus, M. Schienle, R. Thewes, A fully electronic label-free DNA sensor chip. IEEE J. Sens. **7**(4) (2007)

73. C. Hagleitner, D. Lange, A. Hierlemann, O. Brand, H. Baltes, CMOS single-chip gas detection system comprising capacitive, calorimetric and mass-sensitive microsensors. IEEE J. Solid State Circuits, **37**(12) (2002)
74. A. Romani, N. Manaresi, L. Marzocchi, G. Medoro, A. Leonardi, L. Altomare, M. Tartagni, R. Guerrieri, Capacitive sensor array for localization of bioparticles in CMOS lab-on-a-chip. Digest of Technical Papers, IEEE ISSCC Conference, 2004, pp. 224–225
75. S.B. Prakash, P. Abshire, On-chip capacitance sensing for cell monitoring applications. IEEE J. Sens. **7**(3) (2007)
76. Y. Chao, Y. Huang, B.L. Hassler, R.M. Worden, A.J. Mason, Amperometric electrochemical microsystem for a miniaturized protein biosensor array. IEEE Trans. Biomed. Circuits Syst. **3**(3), 160–168 (2009)
77. N. Manaresi, A. Romani, G. Medoro, L. Altomare, A. Leonardi, M. Tartagni, R. Guerrieri. A CMOS chip for individual cell manipulation and detection. IEEE J. Solid State Circuits **38**(12), 2297–2305 (2003)
78. M. Jenkner, M. Tartagni, A. Hierlemann, R. Thewes, Cell-based CMOS sensor and actuator arrays. IEEE J. Solid State Circuits **39**(12), 2431–2437 (2004)
79. G. Medoro, P. Vulto, L. Altomare, M. Abonnenc, A. Romani, M. Tartagni, R. Guerrieri, N. Manaresi, Dielectrophoretic cage-speed separation of bio-particles. Proc. IEEE Sens. **1**, 76–79 (2004)
80. G. Medoro, N. Manaresi, M. Tartagni, R. Guerrieri, CMOS-only sensors and manipulators for microorganisms. Electron Devices Meeting, 2000. IEDM Technical Digest. International, 2000, pp. 415–418
81. M. Tartagni, L. Altomare, R. Guerrieri, A. Fuchs, N. Manaresi, G. Medoro, Microelectronic chips for molecular and cell biology. Sens. Update **13**(1), 155–200 (2003)
82. K.W. Current, K. Yuk, C. McConaghy, P.R.C. Gascoyne, J.A. Schwartz, J.V. Vykoukal, C. Andrews, A High-voltage SOI CMOS exciter chip for a programmable fluidic processor system. IEEE Trans. Biomed. Circuits Syst. **1**(2), 105–115 (2007)
83. H.A. Wake, M.A. Brooke, Low voltage electrophoresis on a CMOS chip. 50th Midwest Symposium on Circuits and Systems (MWSCAS 2007), Aug 2007
84. M. Khorasani, M. Behnam, L. van den Berg, C.J. Backhouse, D.G. Elliott, High-voltage CMOS controller for microfluidics. IEEE Trans. Biomed. Circuits Syst. **3**(2), 89–96 (April 2009)
85. P. Muir, F. Nicholson, M. Jhetam, S. Neogi, J.E. Banatvala, Rapid diagnosis of enterovirus infection by magnetic bead extraction and polymerase chain reaction detection of enterovirus RNA in clinical specimens. J. Clin. Microbiol. **31**(1), 31–38 (1993)
86. T. Aytura, J. Foleyb, M. Anwara, B. Bosera, E. Harrisc, P.R. Beatty, A novel magnetic bead bioassay platform using a microchip-based sensor for infectious disease diagnosis. J. Immunol. Meth. **314**(1), 21–29 (2006)
87. H. Lee, Y. Liu, D. Ham, R. Westervelt, Integrated cell manipulation system – CMOS/microfluidic hybrid. Lab Chip **7**, 331–337 (2007)
88. H. Lee, E. Sun, D. Ham, R. Weissleder, Chip–NMR biosensor for detection and molecular analysis of cells. Nat. Med. **14** (2008)
89. D.E. Schwartz, E. Charbon, K.L. Shepard, A single-photon avalanche diode array for fluorescence lifetime imaging microscopy. IEEE J. Solid State Circuits **43**(11), 2546–2557 (2008)
90. Y. Maruyama, K. Sawada, H. Takao, M. Ishida, A novel filterless fluorescence detection sensor for DNA analysis. IEEE Trans. Electron Devices **53**(3), 553–558 (2006)
91. L. Hartley, K.V.I.S. Kaler, O. Yadid-Pecht, Hybrid integration of an active pixel sensor and microfluidics for cytometry on a chip. IEEE Trans. Circuits Syst. I: Regular Papers **54**(1), 99–110 (2007)
92. M.L. Simpson, G.S. Saylerc, G. Pattersonb, D.E. Nivensc, E.K. Boltonb, J.M. Rochelleb, J.C. Arnottb, B.M. Applegatec, S. Rippc, M.A. Guillornb, An integrated CMOS microluminometer for low-level luminescence sensing in the bioluminescent bioreporter integrated circuit. Sens. Actuator. B Chem. **72**(2), 134–140 (2001)

93. J.G. Bundy, J.L. Wardell, C.D. Campbell, K. Killham, G.I. Paton, Application of bioluminescence-based microbial biosensors to the ecotoxicity assessment of organotins. Lett. Appl. Microbiol. **25**(5), 353–358 (1997)
94. A. El Gamal, H. Eltoukhy, CMOS image sensors. IEEE Mag. Circuits Devices **21**(3), 6–20 (2005)
95. H. Tian, B. Fowler, A.E. Gamal, Analysis of temporal noise in CMOS photodiode active pixel sensor. IEEE J. Solid State Circuits **36**(1), 92–101 (2001)
96. K. Fife, A. El-Gamal, H.S.P. Wong, A multi-aperture image sensor with 0.7 μm pixels in 0.11 μm CMOS technology. IEEE J. Solid State Circuits **43**(12), 2990–3005 (2008)
97. M. Choa, S. Chunga, S.D. Heoa, J. Kua, C. Ban, A simple fluorescent method for detecting mismatched DNAs using a MutS–fluorophore conjugate. Biosens. Bioelectron. **22**(7), 1376–1381 (2007)
98. A. Hierlemann, *Integrated Chemical Microsensor Systems in CMOS Technology* (Springer, New York, 2005)
99. B. Jang, A. Hassibi, Biosensor systems in standard CMOS processes: fact or fiction? IEEE Trans. Ind. Electron. **56**(4), 979–985 (2009)
100. E. Anderson, J. Daniels, H. Yu, T. Lee, N. Pourmand, A label-free CMOS DNA microarray based on charge sensing. IEEE Conference on Instrumentation and Measurement Technology (IMTC), Victoria, BC, May 2008
101. P.M. Levine, G. Ping, R. Levicky, K.L. Shepard, Active CMOS sensor array for electrochemical biomolecular detection. IEEE J. Solid State Circuits **43**(8), 1859–1871 (2008)
102. L. Bandieraa, G. Cellerea, S. Cagninb, A. De Tonia, E. Zanonia, G. Lanfranchib, L. Lorenzelli, A fully electronic sensor for the measurement of cDNA hybridization kinetics. Biosens. Bioelectron. **22**(9), 2108–2114 (2007)
103. Z. Emmanuel Selvanayagam, P. Neuzil, P. Gopalakrishnakonea, U. Sridharb, M. Singh, L.C. Ho. An ISFET-based immunosensor for the detection of β-Bungarotoxin. Biosens. Bioelectron. **17**(9), 821–826 (2002)
104. P.A. Hammond, D. Ali, D.R.S. Cumming, A system-on-chip digital pH meter for use in a wireless diagnostic capsule. IEEE Trans. Biomed. Eng. **52**(4), 687–694 (2005)
105. R.A. Yotter, L.A. Lee, D.M. Wilson, Sensor technologies for monitoring metabolic activity in single cells-part I: optical methods, IEEE Sens. J. **4**(4), 395–411 (2004)
106. R.A. Yotter, D.M. Wilson, Sensor technologies for monitoring metabolic activity in single cells-part II: nonoptical methods and applications. IEEE Sens. J. **4**(4), 412–429 (2004)
107. E. Lauwers, J. Suls, W. Gumbrecht, D. Maes, G. Gielen, W. Sansen, A CMOS multiparameter biochemical microsensor with temperature control and signal interfacing. IEEE J. Solid State Circuits **36**(12), 2030–2038 (2001)
108. P.A. Hammond, D.R.S. Cumming, D. Ali, A single-chip pH sensor fabricated by a conventional CMOS process. IEEE Proc. Sens. (2002)
109. P.K. Chan, D.Y. Chen, A CMOS ISFET interface circuit with dynamic current temperature compensation technique, Trans. Circuits Syst. I: Regular Papers IEEE, **54**(1), 119–129 (2007)
110. A. Hierlemann, H. Baltes, CMOS-based chemical microsensors. Analyst **128**, 15–28 (2003)
111. R.P. Ried, K.E. Sok, D.M. Hong, R.S. Muller, Piezoelectric microphone with on-chip CMOS circuits. IEEE J. Microelectromech. Syst. **2**(3), 111–120 (1993)
112. G. Sauerbrey, Verwendung von Schwingquarzen zur Wägung dünner Schichten und zur Mikrowägung. Zeitschrift für Physik A Hadrons and Nuclei **155**(2) (1959)
113. Q-Sense Inc. http://www.q-sense.com/
114. C. Hagleitner, A. Hierlemann, O. Brand, H. Baltes, Sensor technology CMOS single chip gas detection systems: Part I. Sens. Update **11**(1), 101–155 (2003)
115. SS.S. Leea, R.M. White, Self-excited piezoelectric cantilever oscillators. Sens. Actuator. A: Phys. **52**(1–3), 41–45 (1996)
116. C. Chun-Hao, H. Rong-Zhang, H. Long-Sun, L. Shi-Ming, C. Hsiao-Chin, Y. Yu-Che, L. Yu-Tso, Y. Shih-An, L. Yo-Sheng, W. Yiao-Hong, C. Nai-Kuan, L. Shey-Shi, A wireless bio-MEMS sensor for c-reactive protein detection based on nanomechanics. IEEE Trans. Biomed. Eng. **56**(2), 462–470 (2009)

117. K.-U. Kirstein, Y. Li, M. Zimmermann, C. Vancura, T. Volden, W.H. Song, A. Hierlemannn, J. Lichtenberg, Cantilever-based biosensors in CMOS technology Design. Automation Test Eur. **2**, 1340–1341 (2005)
118. Y. Xiaomei, T. Yaquan, Z. Haitao, T. Li, W. Wei, Design of high-sensitivity cantilever and its monolithic integration with CMOS circuits. IEEE J. Sens. **7**(4), 489–495 (2007)
119. I. Voiculescu, M.E. Zaghloul, R.A. McGill, E.J. Houser, G.K. Fedder, Electrostatically actuated resonant microcantilever beam in CMOS technology for the detection of chemical weapons. IEEE J. Sens. **5**(4), 641–647 (2005)
120. C. Hagleitner, A. Hierlemann, D. Lange, A. Kummer, N. Kerness, O. Brand, H. Baltes, Smart single-chip gas sensor microsystem. Nature **414**, 293–296 (2001)
121. S. Baglio, S. Castorina, N. Savalli, Integrated inductive sensors for the detection of magnetic microparticles. IEEE J. Sens. **5**(3), 372–384 (2005)
122. Z.B. Randjelovic, M. Kayal, R. Popovic, H. Blanchard, Highly sensitive Hall magnetic sensor microsystem in CMOS technology. IEEE J. Solid State Circuits **37**(2), 151–159 (2002)
123. R. Boll, K.J. Overshott, *Magnetic sensors, vol. 5. Sensors: A Comprehensive Survey* (VCH, Weinbeim, 1989)
124. T. Aytura, J. Foleyb, M. Anwara, B. Bosera, E. Harrisc, P.R. Beatty, A novel magnetic bead bioassay platform using a microchip-based sensor for infectious disease diagnosis. J. Immunol. Meth. **314**(1–2), 21–29 (31 July 2006)
125. Aytur, CMOS biosensor for infectious disease detection. Thesis, Electrical Engineering and Computer Sciences University of California, Berkeley, CA, Aug 2007
126. M.A.P. Pertijs, A. Niederkorn, Ma Xu, B. McKillop, A. Bakker, J.H. Huijsing, A CMOS smart temperature sensor with a 3σ inaccuracy of ±0.5°C from −50°C to 120°C. IEEE J. Solid State Circuits **40**(2), 454–461 (2005)
127. M.A.P. Pertijs, K.A.A. Makinwa, J.H. Huijsing, A CMOS smart temperature sensor with a 3σ inaccuracy of ±0.1°C from −55°C to 125°C. IEEE J. Solid State Circuits **40**(12), 2805–2815 (2005)
128. N. Kerness, A. Koll, A. Schaufelbuhl, C. Hagleitner, A. Hierlemann, O. Brand, H. Baltes, N-well based CMOS calorimetric chemical sensors. Thirteenth Annual International Conference on Micro Electro Mechanical Systems (MEMS), Miyazaki, Japan, Jan 2000
129. H. Baltes, O. Paul, O. Brand, Micromachined thermally based CMOS microsensors. Proc. IEEE **86**(8), 1660–1678 (1998)
130. U. Frey, M. Graf, S. Taschini, K.-U. Kirstein, A. Hierlemann, A digital CMOS architecture for a micro-hotplate array. IEEE J. Solid State Circuits **42**(2), 441–450 (2007)
131. M. Graf, D. Barrettino, M. Zimmermann, A. Hierlemann, H. Baltes, S. Hahn, N. Barsan, U. Weimar, CMOS monolithic metal-oxide sensor system comprising a microhotplate and associated circuitry. IEEE J. Sens. **4**(1), 9–16 (2004)
132. B. Derek De Busschere, G.T.A. Kovacs, Portable cell-based biosensor system using integrated CMOS cell-cartridges. Biosens. Bioelectron. **16**(7–8), 543–556 (2001)
133. J.B. Christen, A.G. Andreou, Hybrid silicon/silicone (polydimethylsiloxane) microsystem for cell culture. IEEE International Symposium on Circuits and Systems (ISCAS), Island of Kos, 2006
134. Y. Chingwen, K. Najafi, CMOS interface circuitry for a low-voltage micromachined tunneling accelerometer. J. Microelectromech. Syst. **7**(1), 6–15 (1998)
135. B.V. Amini, F. Ayazi, A 2.5-V 14-bit ΣΔ CMOS SOI capacitive accelerometer. IEEE J. Solid State Circuits **39**(12), 2467–2476 (2004)
136. W.F. Lee, P.K. Chan, A capacitive-based accelerometer IC using injection-nulling switch technique. IEEE Trans. Circuits Syst. I: Regular Papers, **55**(4), 980–989 (2008)
137. B.V. Amini, R. Abdolvand, F. Ayazi, A 4.5-mW closed-loop deltasigma micro-gravity CMOS SOI accelerometer. IEEE J. Solid State Circuits **41**(12), 2983–2991 (2006)
138. L. Benini, C. Guiducci, C. Paulus, Electronic detection of DNA hybridization: toward CMOS microarrays. IEEE Des. Test Comp. **24**(1), 38–48 (2007)
139. A.M. Kummer, A. Hierlemann, Configurable electrodes for capacitive-type sensors and chemical sensors. IEEE J. Sens. **6**(1), 3–10 (2006)

140. E. Ghafar-Zadeh, M. Sawan, D. Therriault, CMOS-based capacitive sensor lab-on-chip: a multidisciplinary approach. Analog Integr. Circuits Signal Process. **59**(1) (2009)
141. E. Ghafar-Zadeh, M. Sawan, D.Therriault, Novel direct-write CMOS-based laboratory-on-chip: design, assembly and experimental results. J. Sens. Actuator A: Phys. **134**(1) (2007)
142. S.B. Prakash, P. Abshire, M. Urdaneta, E. Smela, A CMOS capacitance sensor for cell adhesion characterization 2005. IEEE International Symposium on Circuits and Systems (ISCAS), May 2005
143. S.B. Prakash, P. Abshire, A CMOS capacitance sensor that monitors cell viability. IEEE Conference on Sensors, Irvine, CA, Oct 2005
144. S.B. Prakash, P. Abshire, Tracking cancer cell proliferation on a CMOS capacitance sensor chip. Biosens. Bioelectron. **23**(10), 1449–1457 (2008)
145. A. Hassibi, H. Vikalo, A. Hajimiri, On noise processes and limits of performance in biosensors. J. Appl. Phys. **102**(1–12), 014909 (2007)
146. L. Moreno-Hagelsieb, P.E. Lobert, R. Pampin, D. Bourgeois, J. Remacle, D. Flandre. Sensitive DNA electrical detection based on interdigitated Al/Al2O3 microelectrodes. Sens. Actuator. B: Chem. **98**(2–3), 269–274 (2004)
147. N. Nikkhoo, C. Man, K. Maxwell, P.G. Gulak, A 0.18μm CMOS integrated sensor for the rapid identification of bacteria. Digest of Technical Papers. IEEE International Solid-State Circuits Conference, 2008. ISSCC 2008. 3–7 Feb 2008, pp. 336–617
148. S.M. Radke, E.C. Alocilja, Design and fabrication of a microimpedance biosensor for bacterial detection. IEEE Sens. J. **4**(4), 434–440 (2004)
149. C. Cornila, A. Hierlemann, R. Lenggenhager, P. Malcovati, H. Baltes, G. Noetzel, U. Weimar, W. Göpel, Capacitive sensors in CMOS technology with polymer coating. Sens. Actuator. B: Chem. **25**(1–3), 357–361 (1995)
150. S.M. Radke, E.C. Alocilja, A microfabricated biosensor for detecting foodborne bioterrorism agents. IEEE Sens. J. **5**(4), 744–750 (2005)
151. C. Ou, R. Yuan, Y. Chaia, M. Tanga, R. Chaia, X. Hea, A novel amperometric immunosensor based on layer-by-layer assembly of gold nanoparticles–multi-walled carbon nanotubes-thionine multilayer films on polyelectrolyte surface. Analytica Chimica Acta **603**(2), 205–213 (2007)
152. T. Yina, W. Wei, L. Yanga, X. Gaoa, Y. Gaoa, A novel capacitive immunosensor for transferrin detection based on ultrathin alumina sol–gel-derived films and gold nanoparticles. Sens. Actuator. B: Chem. **117**(1), 286–294 (2006)
153. M. Hnaiena, M.F. Diouanib, S. Helalia, I. Hafaida, W.M. Hassena, N.J. Renaultc, A. Ghramd, A. Abdelghani, Immobilization of specific antibody on SAM functionalized gold electrode for rabies virus detection by electrochemical impedance spectroscopy. Biochem. Eng. J. **39**(3), 443–449 (1 May 2008)
154. Y.Y. Chena, A.C. Shua, H.Y. Hsub, H.Y. Changb, T.R. Yewa, Y.C. Lua, Y.S. Chuanga, Bacteria detection utilizing electrical conductivity. Biosens. Bioelectron. **23**(12), 1856–1861 (2008)
155. A. Sassolas, B.D. Leca-Bouvier, L.J. Blum, DNA biosensors and microarrays. Chem. Rev. **108**, 109–139 (2008)
156. K. Kermana, D. Ozkana, P. Karaa, B. Merica, J.J. Goodingb, M. Ozsoz, Voltammetric determination of DNA hybridization using methylene blue and self-assembled alkanethiol monolayer on gold electrodes. Analytica Chimica Acta **462**(1), 39–47 (2002)
157. H.C. Yoon, M.Y. Hong, H.S. Kim, Affinity biosensor for Avidin using a double functionalized dendrimer monolayer on a gold electrode. Anal. Biochem. **282**, 121–128 (2000)
158. M. Datta, S.A. Merritt, M. Dagenais, Electroless remetallization of aluminum bond pads on CMOS driverchip for flip-chip attachment to vertical cavity surface emitting lasers(VCSEL's). IEEE Trans. Compon. Pack. Technol. **22**(2), 299–306 (1999)
159. L. Berdondini, P.D. Vander-Wa, N.F. De-Rooij, M. Koudelka-Hep, Development of an electroless post-processing technique for depositing gold as electrode material on CMOS devices. Sens. Actuator. B: Chem. **99**(2), 505–510 (2004)
160. F. Hofmann, A.R. Frey, B. Holzapfl, M. Schienle, C. Paulus, P. Schindler-Bauer, R. Thewes, R. Hintsche, E. Nebling, J. Albers, W. Gumbrecht, Passive DNA sensor with gold electrodes

fabricated in a CMOS backend process. 32nd European Conference on Solid-State Device Research Conference, Sept 2002

161. F. Hofmann, A. Frey, B. Holzapfl, M. Schienle, C. Paulus, P. Schindler-Bauer, D. Kuhlmeier, J. Krause, R. Hintsche, E. Nebling, J. Albers, W. Gumbrecht, K. Plehnert, G. Eckstein, R. Thewes, Fully electronic DNA detection on a CMOS chip: device and process issues. Electron Devices Meeting (IEDM '02) Dec 2002, pp. 488–491
162. C. Guiducci, C. Stagni, A. Fischetti, U. Mastromatteo, L. Benini, B. Riccò, Microelectrodes on a silicon chip for label-free capacitive DNA sensing. IEEE Sens. J. **6**(5) (Oct 2006)
163. E. Ghafar-Zadeh, M. Sawan, D. Therriault, Direct-write CMOS-based laboratory-on-chip. Conference and Trade Show NSTI Nanotechnology **2** (2006)
164. E. Ghafer-Zadeh, M. Sawan, A. Shabani, M. Zourob, V. Chodavarapu, Bacteria growth monitoring through a CMOS based capacitive sensor. 14th IEEE International Mixed-Signal, Sensors, and Systems Test Workshop (IMS3TW), Vancouver, 2008
165. F. Di Ingegneria CMOS lab-on-a-chip devices for individual cell biology. Universit`a degli Studi di Bologna Dottorato di Ricerca in Ingegneria Elettronica, Informatica e delle Telecomunicazioni, 2003
166. E. Ghafar-Zadeh, M. Sawan, A charge based capacitive sensor array for Lab-on-chip applications. IEEE J. Sens. **8**(4) (2008)
167. E. Barsoukov, J.R. Macdonald, *Impedance Spectroscopy* (Wiley, New York, 2005)
168. M.I. Prodromidis, Impedimetric immunosensors-A review. Electrochimica Acta (2009)
169. J.G. Guan, Y.Q. Miao, Q.J. Zhang, Impedimetric biosensors. J. Biosci. Bioeng. **97**(4), 219–226 (2004)
170. A. Vermeulen, F. Devlieghere, K. Bernaerts, J. Van Impe, J. Debevere, Growth/no growth models describing the influence of pH, lactic and acetic acid on lactic acid bacteria developed to determine the stability of acidified sauces. Int. J. Food Microbiol. **119**(3), 258–269 (2007)
171. M. Barbaro, A. Bonfiglio, L. Raffo, A. Alessandrini, P. Facci, I. Barák, Fully electronic DNA hybridization detection by a standard CMOS biochip. Sens. Actuator. B: Chem. **118**(1), 41–46 (2006)
172. V. Nanduri, S. Balasubramanian, S. Sista, V.J. Vodyanoy, A.L. Simonian, Highly sensitive phage-based biosensor for the detection of β-galactosidase. Analytica Chimica Acta **589**(2), 166–172 (2007)
173. S. Carrara, V. Bhalla, C. Stagni, L. Benini, A. Ferretti, F. Valle, A. Gallotta, B. Riccò, B. Samorì, Label-free cancer markers detection by capacitance biochip. Sens. Actuator. B: Chem. **136**(1), 163–172 (2009)
174. L. Yao, M. Hajj-Hassan, E. Ghafar-Zadeh, A. Shabani, V. Chodavarapu, M. Zourob, CMOS capacitive sensor system for bacteria detection using phage organisms. IEEE Canadian Conference on Electrical and Computer Engineering (CCECE), Niagara Falls, 2008
175. V.M. Mirsky, M. Riepl, O.S. Wolfbeis, Capacitive monitoring of protein immobilization and antigen-antibody reactions on monomolecular alkylthiol films on gold electrodes. Biosens. Bioelectron. **12**(9), 977–989 (1997)
176. Z. Cheng, Er. Wang, X. Yang, Capacitive detection of glucose using molecularly imprinted polymers. Biosens. Bioelectron. **16**(3), 179–185 (2001)
177. G.J. Yang, J.L. Huang, W.J. Meng, M. Shen, X.A. Jiao, A reusable capacitive immunosensor for detection of *Salmonella* spp. based on grafted ethylene diamine and self-assembled gold nanoparticle monolayers. Anal. Chim. Acta **647**(2), 159–166 (2009)
178. E. Spiller, A. Schöll, R. Alexy, K. Kümmerer, G.A. Urban, A sensitive microsystem as biosensor for cell growth monitoring and antibiotic testing. Sens. Actuator. A: Phys. **130–131**, 312–321 (2006)
179. L. Yang, Y. Li, Detection of viable Salmonella using microelectrode-based capacitance measurement coupled with immunomagnetic separation. J. Microbiol. Meth. **64**, 9–16 (2006)
180. E. Ghafar-Zadeh, M. Sawan, Toward fully integrated CMOS based capacitive sensor for lab-on-chip applications. International Workshop on Medical Measurements and Applications (MeMeA) 2008. IEEE, May 2008, pp. 77–80

181. E. Ghafar-Zadeh, M. Sawan, Towards fully integrated Lab-on-Chip: design, assembly and experimental results. Int. J. Adv. Media Commun. **3**(1), 154–166 (2009)
182. U.A. Nuber, *DNA Microarrays* (Tylor & Francis, Berlin/Germany, 2005)
183. H. Mohamed, L.D. McCurdy, D.H. Szarowski, S. Duva, J.N. Turner, M. Caggana, Development of a rare cell fractionation device: application for cancer detection. IEEE Trans. NanoBiosci. **3**(4), 251–256 (2004)
184. S. Nagrath, L.V. Sequist, S. Maheswaran, D.W. Bell, D. Irimia, L. Ulkus, M.R. Smith, E.L. Kwak, S. Digumarthy, A Muzikansky, P. Ryan, U.J. Balis, R.G. Tompkins, D.A. Haber, M. Toner, Isolation of rare circulating tumour cells in cancer patients by microchip technology. Nature **450** (2007)
185. E.P. Anderson, A CMOS label-free DNA microarray based on charge sensing. Ph.D., Stanford University, 2008
186. S. Parikh, A CMOS imager for DNA detection. M.A.Sc., University of Toronto, 2007
187. X. Chen, L. Jiong, W. Yijin, C. Lu, L. Zuhong, M. Chan. A CMOS-compatible DNA microarray using optical detection together with a highly sensitive nanometallic particle protocol. IEEE Electron Device Lett. **26**(4), 240–242 (2005)
188. T.D. Huang, S. Sorgenfrei, P. Gong, R. Levicky, K.L. Shepard, A 0.18-μm CMOS array sensor for integrated time-resolved fluorescence detection. IEEE J. Solid State Circuit **44**(5), 1644–1654 (2009)
189. G. Sigalov, J. Comer, G. Timp, A. Aksimentiev, Detection of DNA sequences using an alternating electric field in a nanopore capacitor. Nano Lett. **8**(1), 56–63 (2008)
190. M. Barbaro, A. Bonfiglio, L. Raffo, A. Alessandrini, P. Facci, I. BarakBarak. A CMOS, fully integrated sensor for electronic detection of DNA hybridization. IEEE Electron Device Lett. **27**(7) (2006)
191. C. Berggren, P. StaÊlhandske, J. Brundell, G. Johansson, A feasibility study of a capacitive biosensor for direct detection of DNA hybridization. Electroanalysis **11**(3) (1999)
192. C. Guiducci, C. Stagni, G. Zuccheri, A. Bogliolo, L. Benini a, B. Samor`ı b, B. Riccò, DNA detection by integrable electronics. Biosens. Bioelectron. **19**, 781–787 (2004)
193. C. Stagni, C. Guiducci, L. Benini, B. Ricco, S. Carrara, C. Paulus, M. Schienle, R. Thewes, A fully electronic label-free DNA sensor chip. IEEE Sens. J. **7**(4), 577–585 (2007)
194. J.P. Cloareca, J.R. Martina, C. Polychronakosc, I. Lawrenceb, M.F. Lawrenceb, E. Souteyrand, Functionalization of Si/SiO_2 substrates with homooligonucleotides for a DNA biosensor. Sens. Actuator B: Chem. **58**(1), 394–398 (1999)
195. M.L. Yarmush, M. Toner, R. Plonsey, J.D. Bronzino, *Biotechnology for Biomedical Engineers* (CRC, Raton, London, New York, Washington, DC, 2005)
196. P.B. Luppa, L.J. Sokollb, D.W. Chan, Immunosensors – principles and applications to clinical chemistry. Clin. Chim. Acta **314**(1), 1–26 (2001)
197. E. Prusak-Sochaczewski, J.H.T. Luong, Detection of human transferrin by the piezoelectric crystal. Anal. Lett. **23**(2), 183–194 (1990)
198. S.Q. Hua, Z.Y. Wua, Y.M. Zhoua, Z.X. Caoa, G.L. Shen, R.Q. Yu, Capacitive immunosensor for transferrin based on an o-aminobenzenthiol oligomer layer. Anal. Chim. Acta **458**(2), 297–304 (2002)
199. H. Berneya, J. Aldermana, W. Lanea, J.K. Collins, A differential capacitive biosensor using polyethylene glycol to overlay the biolayer. Sens. Actuat. B: Chem. **44**(1), 578–584 (1997)
200. S. Satyanarayanaa, D.T. McCormickb, A. Majumdar, Parylene micro membrane capacitive sensor array for chemical and biological sensing. Sens. Actuat. B: Chem. **115**(1), 494–502 (23 May 2006)
201. J.W. Chunga, S.D. Kima, R. Bernhardtb, J.C. Pyun, Application of SPR biosensor for medical diagnostics of human hepatitis B virus (hHBV). Sens. Actuat. B: Chem. **111**(11), 416–422 (2005)
202. T.H.J. Heutmekersa, M.G.E.G. Bremer, W. Haasnoota, M.W.F. Nielen, A rapid surface plasmon resonance (SPR) biosensor immunoassay for screening of somatotropins in injection preparations. Anal. Chim. Acta **586**(1), 239–245 (2007)

203. C. Berggren, B. Bjarnason, G. Johansson, Capacitive biosensors. Electroanalysis **13**(3), 173–180 (2001)
204. K. Asami, E. Gheorghiu, T. Yonezawa, Real-time monitoring of yeast cell division by dielectric spectroscopy. Biophys. J. **76**(6), 3345–3348 (1999)
205. A.S. Yuwono, P.S. Lammers, Odor pollution in the environment and the detection instrumentation. Intl. J. Sci. Res. Develop. Agric. Eng. **6** (July 2004)
206. E. Ghafar-Zadeh, M. Sawan, D. Therriault, A 0.18-μm CMOS capacitive sensor Lab-on-Chip. Sens. Actuat. A: Phys. **141**(2) (2008)
207. E. Ghafar-Zadeh, D. Therriault, M. Sawan, Programmable three-dimensional microfluidic fabrication by direct-write assembly. NSTI Nanotech, Ecole Polytechnique de Montreal, CA, 2006
208. E. Ghafar-Zadeh, Sawan, D. Therriault, M.A. Miled, Laboratoires-sur-puces: Nouvelle technologie de diagnostic cellulaire et moléculaire. IEEE Can. Rev. **58** (2008)
209. P. Antoniou, J. Hamilton, R. Jain, B. Holloway, B. Koopman, G. Lyberatos, S. A. Svoronos, Effect of temperature and pH on the effective maximum specific growth rate of nitrifying bacteria. Water Res. **24**(1) (1990)
210. EE. Ghafar-Zadeh, M. Sawan, V.P. Chodavarapu, Bacteria growth monitoring through differential CMOS capacitive sensor, TBCAS, Submitted 2009
211. M. Zourob, S. Elwary, A. Turner, *Principles of Bacterial Detection: Biosensors, Recognition Receptors and Microsystems* (Springer, New York, 2007)
212. G. Decher, J.B. Schlenoff, Multilayer thin films. Chapter I of *Polyelectrolyte Multilayers, an Overview* (Wiley-VCH Verlag GmbH, Weinheim, 2002)
213. X. Chen, X. Yan, K.A. Khor, B. Kang, Multilayer assembly of positively charged polyelectrolyte and negatively charged glucose oxidase on a 3D Nafion network for detecting glucose. Biosens. Bioelectron. **22**(12) (2007)
214. S. Zhang. W. Yang, Y. Niu, Y. Li, M. Zhang, C. Sun, Construction of glucose biosensor oxidase onto multilayers of polyelectrolyte/nanoparticles. Anal. Bioanal. Chem. **384**, 736–741 (2006)
215. B. Thierry, F.M. Winnik, Y. Merhi, J. Silver, M. Tabrizian, Bioactive coatings of endovascular stents based on polyelectrolyte multilayers. Biomolecular **15**(7) (2003)
216. F. Durstock, M.F. Rubner, Dielectric properties of polyelectrolyte multilayers. Langumuir **17**(25) (2001)
217. A.L. Hillberg, M. Tabrizian, Biorecognition through layer-by-layer polyelectrolyte assembly: in-situ hybridization on living cells. Biomolecoular **15**(7) (2006)
218. E. Ghafar-Zadeh, M. Sawan, A charge based sigma delta capacitive sensor for ultrathin polyelectrolyte layer detection. Northeast Workshop on Circuits and Systems (NEWCAS) 2008
219. A. Sze, D. Erickson, L. Ren, D. Li, Zeta-potential measurement using the Smoluchowski equation and the slope of the current–time relationship in electroosmotic flow. J. Colloid Interface Sci. **261**(2), 402–410 (2003)
220. E. Ghafar-Zadeh, M. Sawan, A core-CBCM sigma delta capacitive sensor array dedicated to lab-on-chip applications. Sens. Actuat. A: Phys. **144**(2) (2008)
221. Jonathan M. Cooper, Anthony E.G. Cass, *Biosensors* (Oxford University Press, Oxford, 2003)
222. J.V. Sagen, L. Bjørkhaug, J. Molnes, H. Raeder, L. Grevle, O. Søvik, A. Molven, P.R. Njølstad, Diagnostic screening of MODY2/GCK mutations in the Norwegian MODY Registry. Pediatr. Diabetes **9**(5), 442–449 (2008)
223. H.J. Park, S.K. Kim, K. Park, H.K. Lyu, C.S. Lee, S.J. Chung, W.S. Yun, M. Kim, B.H. Chung, An ISFET biosensor for the monitoring of maltose-induced conformational changes in MBP. FEBS Lett. **583**(1) (2009)
224. D.S. Kim, Y.T. Jeong, H.J. Park, J.K. Shin, P. Choi, J.H. Lee, G. Lim, An FET-type charge sensor for highly sensitive detection of DNA sequence. Biosens. Bioelectron. **20**(1) (2004)
225. A. Trifiro, Glucose sensor and uses thereof. U.S. Patent 0232370 A1, 2003
226. R. Lumbroso, N. Naas, L.K. Beitel, M.F. Lawrence, M.A. Trifiro, Novel bioimpedance sensor for glucose recognition. IEEE Conference on Signals, Systems and Electronics (ISSE'07), Montreal, 2007

227. G.B.B. Kristensen, K. Nerhus, G. Thue, S. Sandberg, Standardized evaluation of instruments for self-monitoring of blood glucose by patients and a technologist. Clin. Chem. **50**(6), 1068–1071 (2004)
228. Abotte Inc http://www.abbottdiabetescare.com/
229. A. Salimi, E. Sharifi, A. Noorbakhsh, S. Soltanian, Immobilization of glucose oxidase on electrodeposited nickel oxide nanoparticles: direct electron transfer and electrocatalytic activity. Biosens. Bioelectron. **22**(12), 3146–3153 (2007)
230. J. Wange, Electrochemical glucose biosensors. Chem. Rev. **108**(2), 814–825 (2008)
231. K. Aoki, H. Suzuki, Y. Ishimaru, S. Toyama, Y. Ikariyama, T. Iida, Thermophilic glucokinase-based sensors for the detection of various saccharides and glycosides. Sens. Actuat. B: Chem. **108**(1), 727–732 (2005)
232. S. D'Auria, N. DiCesare, M. Staiano, Z. Gryczynski, M. Rossi, J.R. Lakowicz, A novel fluorescence competitive assay for glucose determinations by using a thermostable glucokinase from the thermophilic microorganism *Bacillus stearothermophilus*. Anal. Biochem. **303**(2) (2002)
233. K. Kamta, M. Mitsuya, T. Nishimura, J. Eiki, Y. Nagata, Structural basis of allosteric regulation of the monimeric allosteric enzyme human glucokinase. Structure **12**(13), 429–438 (2004)
234. J. Molnes, L. Bjorkhaug, O. Sovik, P.R. Njolstad, T. Flatmark, Catalytic activation of human glucokinase by substrate binding-residue contact involved in the binding of D-glucose to the super-open form and conformational transitions. FEBS J. **275**(10), 2467–2481 (2008)
235. E. Ghafar-Zadeh, S.F. Chowdhury, A. Aliakbar, R. Lambrose, V. Chodavarapu, L. Beital, M. Sawan, M. Trifiro, Handheld impedance biosensor system using engineered proteinaceous receptors. Submitted to Biomedical Microdevices, July 2009
236. A. Ulman, Formation and structure of self-assembled monolayers. Chem. Rev. **96**(4), 1533–1554 (1996)
237. F. Tao, Steven L. Bernasek, Understanding odd–even effects in organic self-assembled monolayers. Chem. Rev. **107**(5), 1408–1453 (2007)
238. J.H. Fendler, Chemical self-assembly for electronic applications. Chem. Mat. **13**(2), 3196–3210 (2001)
239. W. Bracke, R. Puers, C.C. Van-Hoof, *Ultra Low Power Capacitive Sensor Interfaces* (Springer, The Netherlands, 2007)
240. E. Ghafar-Zadeh, M. Sawan, A CMOS-Based capacitive censor for laboratory-on-chips: design and experimental results. IEEE International Symposium on Circuits and Systems (ISCAS), New Orleans, 2007
241. N. Yazdi, H. Kulah, K. Najafi, Precision readout circuits for capacitive microaccelerometer. IEEE Proc. Sens. **1**, 28–31 (2004)
242. F.N. Toth, G.C.M. Meijer, A low-cost, smart capacitive position sensor. IEEE Trans. Instrum. Meas. **41**(6), 1041–1044 (1992)
243. B. Rodgers, S. Goenawan, M. Yunus, Y. Kaneko, J. Yoshiike, 16-μA interface circuit for a capacitive flow sensor. IEEE J. Solid State Circuits **33**(12), 2121–2133 (1998)
244. M. Yamada, T. Takebayashi, S. Notoyama, K. Watanabe, A switched-capacitor interface for capacitive pressure sensors. IEEE Trans. Instrum. Meas. **41**(1), 81–86 (1992)
245. S. Xiaojing, H. Matsumoto, K. Murao, A high-accuracy digital readout technique for humidity sensor. IEEE Trans. Instrumen. Meas. **50**(5), 1277–1282 (2001)
246. M. Suster, G. Jun, N. Chaimanonart, W.H. Ko, D.J. Young, A high-performance MEMS capacitive strain sensing system. IEEE J. Microelectromech. Syst. **15**(5), 1069–1077 (2006)
247. Capacitive fluid level sensor Paul Wells, US 5042299, 23 July 1990
248. W. Jiangfeng, G.K. Fedder, L.R. Carley, A low-noise low-offset capacitive sensing amplifier for a 50-μg, Hz monolithic CMOS MEMS accelerometer. IEEE J. Solid State Circuits **39**(5), 722–730 (2004)
249. J.A. Geen, S.J. Sherman, J.F. Chang, S.R. Lewis, Single-chip surface micromachined integrated gyroscope with 50°/h Allan deviation. IEEE J. Solid State Circuits **37**(12), 1860–1866 (2002)
250. M. Tartagni, R. Guerrieri, A fingerprint sensor based on the feedback capacitive sensing scheme. IEEE J. Solid State Circuits **33**(1), 133–142 (1998)

251. N. Manaresi, R. Rambaldi, M. Tartagni, Z.M.K. Vajna, R. Guerrieri, A CMOS-only micro touch pointer. IEEE J. Solid State Circuits **34**(12), 1860–1868 (1999)
252. D. Sylvester, J.C. Chen, H. Chenming, Investigation of interconnect capacitance characterization using charge-based capacitance measurement (CBCM) technique and three-dimensional simulation. IEEE J. Solid State Circuits **33**(3), 449–453 (1998)
253. D. Sylvester, W. Chenming, Analytical modeling and characterization of deep-submicrometer interconnect. Proc. IEEE **89**(5), 634–664 (2001)
254. C. Yao-Wen, C. Hsing-Wen, H. Chung-Hsuan, L. Han-Chao, L. Tao-Cheng, T. Wenchi, J. Ku, L. Chih-Yuan, A novel simple CBCM method free from charge injection-induced errors. IEEE Electron Device Lett. **25**(5), 262–264 (2004)
255. C. Yao-Wen, C. Hsing-Wen, H. Chung-Hsuan, L. Han-Chao, L. Tao-Cheng, T. Wenchi, J. Ku, L. Chih-Yuan, Interconnect capacitance characterization using charge-injection-induced error-free (CIEF) charge-based capacitance measurement (CBCM). IEEE Trans. Semicond. Manuf. **19**(1), 50–56 (2006)
256. R. Bach, B. Davis, R. Laubhan, Improvements to CBCM (charge-based capacitance measurement) for deep submicron CMOS technology. 7th International Symposium on Quality Electronic Design (ISQED '06), San Jose, CA, March 2006
257. Z. Hui, K. Raseong, A. Paul, M. Luisier, G. Klimeck, M. Fa-Jun, S.C. Rustagi, G.S. Samudra, N. Singh, L. Guo-Qiang, K. Dim-Lee, Characterization and modeling of subfemtofarad nanowire capacitance using the CBCM technique. IEEE Electron Device Lett. **30**(5), 526–528 (2009)
258. C. Yao-Wen, C. Hsing-Wen, L. Tao-Cheng, K. Ya-Chin, T. Wenchi, K.Y.H. Joseph, L. Chih-Yuan, Charge-based capacitance measurement for bias-dependent capacitance. IEEE Electron Device Lett. **27**(5), 390–392 (2006)
259. B. Sell, A. Avellan, W.H. Krautschneider, Charge-based capacitance measurements (CBCM) on MOS devices. IEEE Trans. Device Mat. Reliab. **2**(1), 9–12 (2002)
260. C. Guiducci, C. Stagni, G. Zuccheri, A. Bogliolo, L. Benini, B. Samori, B. Ricco, A biosensor for direct detection of DNA sequences based on capacitance measurements. 32nd European Solid-State Device Research Conference, 2002
261. S.B. Prakash, P. Abshire, A fully differential rail-to-rail CMOS capacitance sensor with floating-gate trimming for mismatch compensation. IEEE Trans. Circuits Syst. I: Regular Papers **56**(5), 975–986 (2009)
262. E. Ghafer-Zadeh, M. Sawan, Toward fully integrated CMOS based capacitive sensor lab-on-chip. IEEE International Workshop on Medical Measurements and Applications (MeMeA), Ottawa, 2008
263. E. Ghafar-Zadeh, M. Sawan, CMOS based capacitive detection lab-on-chip: design and experimental results. IEEE Custom Integrated Circuits Conference (CICC), San Jose, CA, 2007
264. E. Ghafer-Zadeh, M. Sawan, A hybrid IC/microfluidic sensor for lab-on-chips. 4th Canadian Workshop on MEMS and Microfluidic, Montreal, 2007
265. E. Ghafar-Zadeh, M. Sawan, A highly linear capacitive sensor circuit based on a novel CBCM method. 3th International IEEE Northeast Workshop on Circuits and Systems (NEWCAS), Quebec City, 2005
266. E. Ghafar-Zadeh, M. Sawan, A high accuracy differential capacitive circuit for bioparticles sensing applications. IEEE International Midwest Symposium on Circuits and Systems (MWSCAS), Montreal (2007)
267. I. Evans, T. York, Microelectronic capacitance transducer for particle detection. IEEE Sens. J. **4**(3), 364–372 (2004)
268. T.A. York, G. Evansa, Z. Pokusevski, T. Dyakowskib, Particle detection using an integrated capacitance sensor. Sens. Actuators A: Phys. **92**(1), 74–81 (2001)
269. E. Ghafar-Zadeh, M. Sawan, Charge based capacitive sensor array for CMOS based laboratory-on-chip applications. IEEE Conference on Sensors (5th), Oct 2007
270. E. Ghafar-Zadeh, M. Sawan, D. Therriault, Microelectrofluidic packaging for CMOS based laboratory-on-chips. International Electronics Packaging Symposium, Niskayuna, 2006

271. S.R. Norsworthy, R. Schreier, G.C. Temes. Delta-sigma data converters: theory, design, and simulation. Wiley-IEEE Press, (1996)
272. M. Yin, F.O. Eynde, W. Sansen, A high-speed CMOS comparator with 8-b resolution. IEEE J. Solid State Circuits **27**(2) (1992)
273. S. Hein, K. Ibraham, A. Zakhor, New properties of sigma-delta modulators with DC inputs. IEEE Trans. Commun. **40**(8) (1992)
274. L.G. McIlrath, A robust O (N log n) algorithm for optimal decoding of first-order Σ-Δ sequences. IEEE Trans. Signal Process. **50**(8) (2002)
275. F. Dachselt, M. Gotz, Rational cycle decoding algorithm for the first-order delta-sigma modulator. IEEE International Symposium on Circuits and Systems (ISCAS), May 2001
276. F. Dachselt and S. Quitz, Structure and information content in sequence from the single-loop sigma-delta modulator with dc inpu. IEEE International Symposium on Circuits and Systems ISCAS, **4**, 685–688 (May 2004)
277. M.A. Miled, E. Ghafar-Zadeh, M. Sawan, Fast decoding algorithm for first order DC-input sigma-delta modulators. 50th Midwest Symposium on Circuits and Systems (MWSCAS), 2007
278. M.A. Miled, M. Sawam, E. Ghafar-Zadeh, A dynamics decoder for first-order sigma–delta modulators dedicated to lab-on-chip applications. IEEE Trans. Signal Process. **57**(10), 4076–4084 (2009)
279. S. Hardt, F. Schönfeld, Microfluidic Technologies for Miniaturized Analysis Systems (Springer, New York, 2007)
280. M. Agirregabiria, F.J. Blanco, J. Berganzo, M.T. Arroyo, A. Fullaondo, K. Mayora, J.M. Ruano-López, Fabrication of SU-8 multilayer microstructures based on successive CMOS compatible adhesive bonding and releasing steps. Lab Chip **5**, 545–552 (2005)
281. N.H. Tea, V. Milanovic, C.A. Zincke, J.S. Suehle, M. Gaitan, M.E. Zaghloul, J. Geist, Hybrid postprocessing etching for CMOS-compatible MEMS. J. Microelectromechan. Syst. **6**(4), 363–372 (1997)
282. P. Zheng-chun, L. Zhong-geng, M. Tondra, L. Chang-geng, M. Zhang, K. Lian, J. Goettert, J. Hormes, CMOS compatible integration of three-dimensional microfluidic systems based on low-temperature transfer of SU-8 films. J. Microelectromech. Syst. **15**(3) (2006)
283. F.J. Blanco, M. Agirregabiria, J. Garcia, J. Berganzo, M. Tijero, M.T. Arroyo, J.M. Ruano, I. Aramburu, Kepa Mayora, Novel three-dimensional embedded SU-8 microchannels fabricated using a low temperature full wafer adhesive bonding. Micromech. Microeng. **14** (2004)
284. G. Kaltsas, D.N. Pagonis, A.G. Nassiopoulou, Planar CMOS compatible process for the fabrication of buried microchannels in silicon, using porous-silicon technology. J. Microelectromech. Syst. **12**(6), 863–872 (2003)
285. A. Rasmussen, M.E. Zaghloul, CMOS microfluidic fabrication technology for biomedicalapplications. 42nd Midwest Symposium on Circuits and Systems **2**, 791–794 (1999)
286. E. Ghafar-Zadeh, M. Sawan, D. Therriault, A New approach for the integration of microfluidic structures to microelectronic devices. CMC Microsystems, 4th Canadian Workshop on MEMS (2006)
287. P.F. Man, D.K. Jones, C.H. Mastrangelo, Microfluidic plastic capillaries on silicon substrates: a new inexpensive technology for bioanalysis chips. IEEE Micro Electro Mechanical Systems (MEMS), 1997
288. M.A. Burns, B.N. Johnson, S.N. Brahmasandra, K. Handique, J.R. Webster, M. Krishnan, T.S. Sammarco, P.M. Man, D. Jones, D. Heldsinger, C.H. Mastrangelo, D.T. Burke. An integrated nanoliter DNA analysis device. Science **282**(5388), 484–487 (16 October 1998)
289. A. Rasmussen, M. Gaitan, L.E. Locascio, M.E. Zaghloul, Fabrication techniques to realize CMOS-compatible microfluidicmicrochannels. J. Microelectromech. Syst. **10**(2) (2001)
290. A. Rasmussen, Implementation and modeling of microfluidic components realized using CMOS technology. Angela, D.Sc., George Washington University, 2002
291. H. Lee, D. Ham, R.M. Westervelt, CMOS/microfluidic hybrid systems. Chapter III in *CMOS Biotechnology* (Springer Science+Business Media, LLC (2007)
292. H. Lee, Y. Liu, R.M. Westervelt, D. Ham, IC/microfluidic hybrid system for magnetic manipulation of biological cells. IEEE J. Solid State Circuits **41**(6) (2006)

293. I. Chartier, C. Bory, A. Fuchs, D. Freida, N. Manaresi, M. Ruty, J. Bablet, L. Fulbert, Fabrication of hybrid plastic-silicon micro-fluidic devices for invidual cell manipulation by dielectrophoresis. Proc. SPIE **5345** (2004)
294. P. Sethu, C.H. Mastrangelo, Cast epoxy-based microfluidic systems and their application in biotechnology. Sens. Actuat. B: Chem. **98**(2), 337–346 (2004)
295. P. Sethu, C.H. Mastrangelo, Polyethylene glycol (PEG)-based actuator for nozzle–diffuser pumps in plastic microfluidic systems. Sens. Actuat. A: Phys. **104**(3), 283–289 (2003)
296. P. Vulto, N. Glade, L. Altomare, J. Bablet, L. Tin, G. Del-Medoro, I. Chartier, N. Manaresi, M. Tartagni, R. Guerrieri, Microfluidic channel fabrication in dry film resist for production and prototyping of hybrid chips. J. Lab chip **5** (2005)
297. J.H. Song, M.J. Edirisinghe, J.R.G. Evans, Formulation and multilayer jet printing of ceramic inks. J. Am. Ceram. Soc. **82**(12) (1999)
298. S.L. Morissette, J.A. Lewis, P.G. Clem, J. Cesarano, D.B. Dimos, Direct-write fabrication of Pb(Nb,Zr,Ti)O_3 devices: influence of paste rheology on print morphology and component properties, J. Am. Ceram. Soc. **84**(11) (2001)
299. K.A.M. Seerden, N. Reis, J.R.G. Evans, P.S. Grant, J.W. Halloran, B. Derby, Ink-jet printing of wax-based alumina suspensions. J. Am. Ceram. Soc. **84**(11) (2001)
300. J.A. Lewis, Direct-write assembly of ceramics from colloidal inks, Current. Opin. Solid State Mater. Sci. **6**(3) (2002)
301. M. Xu, G.M. Gratson, E.B. Duoss, R.F. Shepherd, J.A. Lewis, Biomimetic silicification of 3D polyamine-rich scaffolds assembled by direct ink writing. Soft Matter **16**(9) (2006)
302. G.M. Gratson, F. Garcia-Santamaria, V. Lousse, M. Xu, S. Fan, J.A. Lewis, P.V. Braun, Direct-write assembly of three-dimensional photonic crystals: conversion of polymer scaffolds to silicon hollow-woodpile structures. Adv. Mater. **18**(4) (2006)
303. J.L. Simon, S. Michna, J.A. Lewis, E.D. Rekow, V.P. Thompson, J.E. Smay, A. Yampolsky, J.R. Parsons, J.L. Ricci, In vivo bone response to 3D periodic hydroxyapatite scaffolds assembled by direct ink writing. J. Biomed. Mater. Res. A **26**(28) (2007)
304. J.G. Dellinger, J. Cesarano 3rd, RD Jamison Robotic deposition of model hydroxyapatite scaffolds with multiple architectures and multiscale porosity for bone tissue engineering. J. Biomed. Mater. Res. A **82**(2) (2007)
305. D. Therriault, S.R. White, J.A. Lewis, Chaotic mixing in three-dimensional microvascular networks. Nat. Mater. **2**(4) (2003)
306. E. Ghafar-Zadeh, M. Sawan, D. Therriault, A microfluidic packaging technique for lab-on-chip applications. EEE Techol. Adv. Pack. **32**(2) (2009)
307. M. Hajj-Hassan, T. Gonzalez, E. Ghafar-Zadeh, H. Djeghelian, V. Chodavarapu, M. Andrews, D. Therriault, Direct-dispense polymeric waveguides platform for optical chemical sensors. Sensors **8**(12) (2008)
308. M. Kuhn, T. Napporn, M. Meunier, S. Vengallatore, D. Therriault, Direct-write microfabrication of single-chamber micro solid oxide fuel cells. J. Micromech. Microeng. **18**(1) (2008)
309. M. Kuhn, T. Napporn, M. Meunier, D. Therriault, S. Vengallatore, Direct-write microfabrication of single-chamber solid oxide fuel cells with interdigitated electrodes. Mater. Res. Soc. Symp. Proc. **972** (2007)
310. M. Kuhn, T. Napporn, M. Meunier, D. Therriault, S. Vengallatore, Fabrication and testing of coplanar single-chamber micro solid oxide fuel cells with geometrically complex electrodes. J. Power Source. **177**(1) (2008)
311. T. Hibino, H. Iwahara, Simplification of solid oxide fuel cell systems using partial oxidation of methane. Chem. Lett. **7** (1993)
312. M. Nagao, M. Yano, K. Okamoto, A. Tomita, Y. Uchiyama, N. Uchiyama, T. Hibino, A single-chamber sofc stack: energy recovery from engine exhaust. Fuel Cell. **8**(5) (2008)
313. Z. Shao, S.M. Haile, J. Ahn, P.D. Ronney, Z. Zhan, S.A. Barnett, A thermally self-sustained micro solid-oxide fuel-cell stack with high power density. Nature **435**(7043) (2005)
314. B. Morel, R. Roberge, S. Savoie, T. W. Napporn, M. Meunier, An experimental evaluation of the temperature gradient in solid oxide fuel cells. Electrochem. Solid State Lett. **10**(2) (2007)

315. D. Therriault, R.F. Shepherd, S.R. White, J.A. Lewis, Fugitive inks for direct-write assembly of 3-D microvascular networks. Adv. Mater. **17**(4) (2007)
316. B.R. Flachsbart, K. Wong, J.M. Iannacone, E.N. Abante, R.L. Vlach, P.A. Rauchfuss, P.W. Bohn, J.V. Sweedler, M.A. Shannon, Design and fabrication of a multilayered polymer microfluidic chip with nanofluidic interconnects via adhesive contact printing. Lab Chip **6**, 667–674 (2006)
317. X.B. Chen, W.J. Zhang, G. Schoenau, B. Surgenor, Off-line control of time-pressure dispensing processes for electronics packaging. IEEE Trans. Electron. Pack. Manuf. **26**(4) (2003)
318. X.B. Chen, H. Ke, Effects of fluid properties on dispensing processes for electronics packaging. IEEE Trans. Electron. Pack. Manuf. **29**(2) (2006)
319. R. Rao, A. Morales, K. Kracik, J.A. Lewis, Micro-fabricated deposition nozzles for direct write assembly of 3D periodic structures. J. Adv. Mater. **17**(3) (2005)
320. E. Ghafar-Zadeh, M. Sawan, D. Therriault, Direct-write fabrication of microchannel in epoxy resin. ASME Mechanical Engineering Congress and Exposition(IMECE), Orlando, FL, 2005
321. H. Becker, C. Gärtner, Polymer microfabrication technologies for microfluidic systems. Anal. Bioanal. Chem. **39**(1) (2008)
322. R.H. Liu, Q. Yu, D.J. Beebe, Fabrication and characterization of hydrogel-based microvalves. J. Microelectromech. Syst. **11**(1) (2002)
323. T. Miyata, N. Asami, T. Uragami, A reversibly antigen-responsive hydrogel. Nature **399**(766) (1999)
324. Liang Dong, Hongrui Jiang, Autonomous microfluidics with stimuli-responsive hydrogels. Soft Matter. **3**(10) (2007)
325. J. Wang, Z. Chen, M. Mauk, K. Sheng Hong, M. Li, S. Yang, H.H. Baul, Self-actuated, thermo-responsive hydrogel valves for labon a chip. Biomed. Device. **7**(4) (2005)
326. E. Ghafar-Zadeh, M. Sawan, V. Chodavarapu, A direct-write microfluidic fabrication process for CMOS-based Lab-on-Chip applications. Microelectron. Eng. **86**(10) (2009)
327. M.L. Berre, G. Pandraud, P. Morfouli, M. Lallemand, The performance of micro heat pipes measured by integrated sensors. J. Micromech. Microeng. **16** (2006)
328. R. Bey-Oueslati, S. Martel, D. Therriaul, Micro heat pipe fabrication: high performance deposition platform for electronic industry. International Workshop on Microfactories, 2006
329. P. de la Fuentea, J.A. Etxeberriaa, J. Berganzob, J.M. Ruano-Lópezb, M.T. Arroyob, E. Castañoa, F.J. Graciaa, End-fire coupling of a SU-8 waveguide to a silicon mesa photodiode: Integrability in an optical analysis microsystem. Sens. Actuat. A: Phys. **123–124**, 313–318 (2005)
330. Z. Ye, Q. Lou, J. Dong, Y. Wei, L. Lin, Compact continuous-wave blue lasers by direct frequency doubling of laser diodes with periodically poled lithium niobate waveguide crystals. Opt. Lett. **30**(1), 73–74 (2005)
331. H. Huanga, Z. Liu, X. Yang. Application of electrochemical impedance spectroscopy for monitoring allergen–antibody reactions using gold nanoparticle-based biomolecular immobilization method. Anal. Biochem. **356**(2), 208–214 (2006)
332. F. Lucarelli, G. Marrazza, A.P.F. Turner, M. Mascini, Carbon and gold electrodes as electrochemical transducers for DNA hybridisation sensors. Biosens. Bioelectron. **19**(6), 515–530 (2004)
333. Abaxis http://www.abaxis.com/piccologg/
334. K. Bourzac, Rapid TB detector: an ultrasensitive test can spot bacteria in a half hour. MIT Technology Review Tuesday, Aug 2004
335. H.-Y.K. Estes, C.R. Duncan, A.G. Wade, B.D. Cleary, F.C. Lloyd, C.R. Ellis, W.R. Jr. L.S. Powers. Real-time detection of microbial contamination. IEEE Eng. Med. Biol. **23**(1) (2004)
336. E. Spiller, A. Schöll, R. Alexy, K. Kümmerer, G.A. Urban, A sensitive microsystem as biosensor for cell growth monitoring and antibiotic testing. Sens. Actuat. A: Phys. **130–131**(14) (2006)
337. L.L. Hause, R.A. Komorowski, F. Gayon, Electrode and electrolyte impedance in the detection of bacterial growth. IEEE Trans. Biomed. Eng. **28**(5) (1981)
338. L.Yang, Electrical impedance spectroscopy for detection of bacterial cells in suspensions using interdigitated microelectrodes. Talanta **74**(5) (2008)

339. V. Velusamy, K. Arshak, O. Korostynska, K. Oliwa, C. Adley, Design of a real time biorecognition system to detect foodborne pathogens-DNA biosensor. IEEE Sensors Applications Symposium (SAS), 2009
340. O. Korostynska, K. Arshak, A. Arous, S. Fitzpatrick, Instant glucose sensing using the AD5933 impedance converter. E-MRS Symposium Fall Meeting, 2009
341. L. Majer, V. Stopjaková, E. Vavrinský, Sensitive and accurate measurement environment for continuous biomedical monitoring using microelectrodes. 11th IEEE Workshop on Design and Diagnostics of Electronic Circuits and Systems (DDECS), 2008
342. Microlyne http://www.micralyne.com
343. Micronit http://www.micronit.com/

Index

MIX
Papier aus verantwortungsvollen Quellen
Paper from responsible sources
FSC® C105338

If you have any concerns about our products, you can contact us on
ProductSafety@springernature.com

In case Publisher is established outside the EU, the EU authorized representative is:
Springer Nature Customer Service Center GmbH
Europaplatz 3, 69115 Heidelberg, Germany

Printed by Libri Plureos GmbH
in Hamburg, Germany